Sustainable Digital Technologies

This book is a reference on digital technology and its impact on sustainability, providing insight into sustainable practices globally. It focuses on the critical practices leading to sustainable initiatives among various organizations, IT infrastructure, communities, and government compliance. The book describes the green computing paradigms and the impact of a circular economy with a focus on sustainable practices in a post-pandemic world.

Sustainable Digital Technologies: Trends, Impacts, and Assessments discusses the critical factors leading to sustainable initiatives in a global economy. It highlights the impact of digital technology and Industry 4.0 in today's world. The book focuses on the role, responsibility, and the effect of the Internet of Things for digital sustainability and practices. It describes implementation strategies for green cloud computing and presents additional strategies for sustainable practices in a post-pandemic world.

This publication is designed for use by technology development academicians, data scientists, industrial professionals, researchers, and students interested in uncovering the latest innovations in the field and the current research on problem-oriented processing techniques in sustainable and evolutionary computing applications with reduced energy channelization.

Sustainable Digital Technologies

Trends, Impacts, and Assessments

Edited by
Vandana Sharma, Balamurugan Balusamy,
Munish Sabharwal, and Mariya Ouaissa

CRC Press
Taylor & Francis Group
Boca Raton London New York

CRC Press is an imprint of the
Taylor & Francis Group, an **informa** business

Designed cover image: Shutterstock

First edition published 2024
by CRC Press
6000 Broken Sound Parkway NW, Suite 300, Boca Raton, FL 33487-2742

and by CRC Press
4 Park Square, Milton Park, Abingdon, Oxon, OX14 4RN

CRC Press is an imprint of Taylor & Francis Group, LLC

Library of Congress Cataloging-in-Publication Data

Library of Congress Cataloging-in-Publication Data
Names: Sharma, Vandana, editor. | Balusamy, Balamurugan, editor. | Sabharwal, Munish, editor. | Ouaissa, Mariya, editor.
Title: Sustainable digital technologies : trends, impacts, and assessments / edited by Vandana Sharma, Balamurugan Balusamy, Munish Sabharwal, Mariya Ouaissa.
Description: First edition. | Boca Raton : CRC Press, 2024. | Includes bibliographical references and index.
Identifiers: LCCN 2023010579 (print) | LCCN 2023010580 (ebook) | ISBN 9781032387666 (hbk) | ISBN 9781032390833 (pbk) | ISBN 9781003348313 (ebk)
Subjects: LCSH: Sustainable engineering—Data processing. | Digital communications—Environmental aspects. | Computer science—Environmental aspects.
Classification: LCC TA170 .S8685 2024 (print) | LCC TA170 (ebook) | DDC 004.6028/6--dc23/eng/20230330
LC record available at https://lccn.loc.gov/2023010579
LC ebook record available at https://lccn.loc.gov/2023010580

ISBN: 978-1-032-38766-6 (hbk)
ISBN: 978-1-032-39083-3 (pbk)
ISBN: 978-1-003-34831-3 (ebk)

DOI: 10.1201/9781003348313

Typeset in Times
by Deanta Global Publishing Services, Chennai, India

Contents

Editors....vii
List of Contributors....ix

Chapter 1 Critical Factors Leading to Sustainable Initiatives in the Global Market....1

Anamika Singh, Vandana Sharma, Rajesh Kumar Dhanaraj, Balamurugan Balusamy, Munish Sabharwal, and Mariya Ouaissa

Chapter 2 Patents and Innovations for Digital Sustainability....27

Sonu Kumar Jha, Udit Narayan, Rahul Kumar, Vandana Sharma, Naman Mishra, and Yogita Rawat

Chapter 3 Adoption of Sustainable Digital Technologies in Industry 4.0....51

Naman Mishra, Vandana Sharma, and Achyut Shankar

Chapter 4 Industry 4.0: Features Adopting Digital Technologies in the Oil and Gas Industry....69

Nasurudeen Ahamed N., Amreen Ayesha, and P. Karthikeyan

Chapter 5 Effects of Contaminated Ocean Water on the Ecosystem: A Review Using Machine Learning....91

Samiksha Sharma, Aman Verma, Kajal K., and Siva Rama Krishnan Somayaji

Chapter 6 Trends and Challenges in Green Computing....109

Priti Rai and Yogita Rawat

Chapter 7 Implementation Strategies for Green Computing....135

Kuldeep Singh Kaswan, Jagjit Singh Dhatterwal, Santar Pal Singh, Vandana Sharma, and Balamurugan Balusamy

Chapter 8 Strategies for Sustainable Practices in a Post-Pandemic World....149

Jagjit Singh Dhatterwal, Kuldeep Singh Kaswan, Sanjay Kumar, Vandana Sharma, and Balamurugan Balusamy

Chapter 9 Strategies for Sustainable Supply Chains in a Circular Economy: Bibliometric Analysis and an Overview 161

Laxmi Pandit Vishwakarma and Rajesh Kr Singh

Chapter 10 Research Potentials and Future Trends of Digital Sustainability 187

Shilpy Verma, Vandana Sharma, Vinay Kukreja, Achyut Shankar, and Ahmed A. Elngar

Index 205

Editors

Vandana Sharma is an Associate Professor at Amity Institute of Information Technology, Amity University Noida Campus, India. She has served NAAC-accredited A++ institutions like Guru Gobind Singh Indraprastha University, Delhi and Galgotias University, Greater Noida Campus. Dr. Sharma has 14+ years of teaching experience at the postgraduate level and has guided 100+ student dissertations. She obtained her doctoral, masters of technology and bachelor of electronic commerce degrees from Guru Gobind Singh Indraprastha University, a Delhi state university in India. She is a Senior Member of IEEE, member of Women in Engineering Society, and member of IEEE Consumer Technology Society popularly known as CTSoc. As a keen researcher, she has published 50+ research papers in SCI and Scopus Indexed international journals and conferences. She has contributed voluntarily as a keynote speaker, session chair, Reviewer and Technical Program Committee (TPC) member for reputed International Journals and IEEE Conferences and has presented her work across India and abroad. Currently, she is a post-doctoral fellow at Lincoln University College, Malaysia. Also, she is honorary Adjunct Professor and Senior Fellow of Scientific Innovation Research Group (SIRG) at Beni Suef University, Egypt.

Her primary areas of interest include artificial intelligence, blockchain technology and the Internet of Things (IoT).

Balamurugan Balusamy has served as a professor and has 14 years of experience with VIT University, Vellore. He completed his bachelor's, master's, and doctorate at top premier institutions in India. He has published 30-plus books on various technologies and visited 15-plus countries teaching his technical course. He has several top-notch conferences on his resume and has been published in over 150 quality journals, conferences, and book chapters. Dr. Balusamy serves on the advisory committee for several start-ups and forums, and does consultant work on industrial IoT. He has given over 175 talks at various events and symposiums, and is currently working as an associate dean at Shiv Nadar University, India. Prior to this assignment, he was a professor at Galgotias University; his research interests are blockchain and IoT.

Munish Sabharwal has a doctorate in computer science and a doctorate in management information systems, and has contributed more than 22 years to teaching education management and research as well as software development. He is currently spearheading efforts as dean and professor at Galgotias University, Greater Noida, India. His research interests include data science, artificial intelligence, machine learning, biometrics, and e-banking.

Mariya Ouaissa is a professor/trainer, researcher, and practitioner with industry and academic experience. She earned a doctorate in 2019 in computer science and networks at the Laboratory of Modelisation of Mathematics and Computer Science from ENSAM, Moulay Ismail University, Meknes, Morocco. She is a networks and telecoms engineer, who graduated in 2013 from the National School of Applied Sciences Khouribga. Her main research topics are Internet of Things (IoT), machine to machine (M2M), device to device (D2D), wireless sensor networks (WSN), cellular networks, and vehicular networks.

List of Contributors

Amreen Ayesha
Presidency University
India

Balamurugan Balusamy
Shiv Nadar Institute of Eminence
Shiv Nadar University
Delhi NCR Campus, India

Rajesh Kumar Dhanaraj
Symbiosis International (Deemed University)
Pune, India

Jagjit Singh Dhatterwal
Koneru Lakshmaiah Education Foundation
India

Ahmed A. Elngar
Faculty of Computers and Artificial Intelligence
Beni-Suef University
Beni-Suef, Egypt

Kajal K.
Vellore Institute of Technology
India

P. Karthikeyan
National Ching Cheng University
Taiwan

Kuldeep Singh Kaswan
Galgotias University
India

Rahul Kumar
Galgotias University
India

Sanjay Kumar
KCC Institute of Technology & Management
India

Sonu Kumar Jha
Galgotias University
India

Naman Mishra
FS University
India

Nasurudeen Ahamed N.
Presidency University
India

Udit Narayan
Galgotias University
India

Mariya Ouaissa
Moulay Ismail University
Morocco

Priti Rai
GL Bajaj Institute of Management
Greater Noida, India

Yogita Rawat
ITM Business School
India

Munish Sabharwal
Galgotias University
India

Achyut Shankar
University of Warwick
United Kingdom

Samiksha Sharma
Vellore Institute of Technology
India

Vandana Sharma
Computer Science Department
CHRIST (Deemed to be University)
Delhi-NCR, India

Anamika Singh
Galgotias University
India

Rajesh Kr Singh
Management Development Institute
Gurgaon, India

Santar Pal Singh
Rashtrakavi Ramdhari Singh Dinkar College of Engineering
India

Siva Rama Krishnan Somayaji
Vellore Institute of Technology
India

Aman Verma
Vellore Institute of Technology
India

Shilpy Verma
Amity University
Noida Campus, India

Laxmi Pandit Vishwakarma
Management Development Institute
Gurgaon, India

1 Critical Factors Leading to Sustainable Initiatives in the Global Market

Anamika Singh, Vandana Sharma,
Rajesh Kumar Dhanaraj, Balamurugan Balusamy,
Munish Sabharwal, and Mariya Ouaissa

CONTENTS

1.1 Introduction to Sustainable Development 1
1.1.1 Definition 4
1.1.2 History of Sustainable Development 5
1.1.3 Various Definitions of Sustainable Development 5
1.2 Literature Review 7
1.3 Components of Sustainable Development 8
1.3.1 Economic Sustainability 9
1.3.2 Social Sustainability 10
1.3.3 Ecological Sustainability 11
1.4 Implementation Progress of Sustainable Development 11
1.5 Digital Sustainability 11
1.6 Critical Factors Leading to Sustainable Initiatives in the Global Market 13
1.7 Sustainable Development Goals (SDGs) 15
1.8 Sustainable Initiatives in the IT Sector around the World 19
1.9 Role of Artificial Intelligence in Sustainable Development 21
1.10 Conclusion 23
Bibliography 23

1.1 INTRODUCTION TO SUSTAINABLE DEVELOPMENT

The concept of sustainable development originates from the word *sustainable* of which the objective is to avoid the reduction of the natural system in contemplation of maintaining and managing the ecological balance. Sustainable development focuses on building a sustainable, comprehensive, and strong future without hampering the natural resources for people and vegetation. Sustainable development is a global issue (Ahmed et al., 2022). The term *sustainable development* was popularized in 1987 by the United Nations (UN) Commission of Environment and Development through the Brundtland Report. By the 2000s, globalization had reached almost every corner

DOI: 10.1201/9781003348313-1

of the world, making humans more aware of the environmental crisis and damage to the natural system (Ali et al., 2022a). The concern for the environmental crisis and the disbalance in the natural system/natural resources has brought forth the need for a new and efficient agenda or model that avoids more damage to the natural system and maintains an adequate ecological balance. In the history of sustainable development, the year 2015 was a landmark for the shaping of international policies when a set 17 Sustainable Development Goals (SDGs) was established and the 2030 Agenda of Sustainable Development was adopted by the UN (Ali et al., 2022b).

The most extensively acquired definition for sustainable development was proposed by the Brundtland Commission in its "Our Common Future" report (1987). The definition is stated as the development that fulfils the desires of the present without compromising the needs of the future generation. Sustainability has become a promising practice, irrespective of the working environment and underlying implementing conditions (Bank, 2013). There has been notable growth in the interest and acceptance of sustainability and sustainable development in academics, policies, and in all working scenarios. Sustainability and sustainable development are two broad concepts, but hold technically different meanings. Sustainability could be defined as an approach or the principles that integrate and bring together all the aspects of development with the objective to achieve sustainable development (Basheer et al., 2023).

Sustainability and sustainable development cover a very broad sense depending on the underlying domain. The significance of sustainable development is widely adopted in academics for four major reasons:

- Sustainable development is contemplated as the UN's end goal for the planet.
- Sustainable living with a sustainable planet for every generation is promoted by sustainable development.
- All other developments aim to achieve sustainable development, hence sustainable development is considered as embracing.
- Long-lasting socioeconomic advantages for humans as well as the environment are expected to be achieved by sustainable development.

Sustainable development is defined as the effective management of resources and the natural system in a way to meet present-generation needs and the needs of the future generation as well. Allocating and assigning required resources to economic as well as non-economic bodies is determined by the sustainable development (David, 2018).

Several themes in sustainability and sustainable development have emerged such as the compelling reasons for sustainability and sustainable development, promotion of sustainable development by building reconfiguration and revolution, varying approaches related to sustainable development, sustainable development practices for specific countries, and achieving sustainable development with responsibilities toward the environment and growth of economy. There have been many advances made in the field of sustainable development, but issues still persist that should be addressed (Duran et al., 2015).

The National Sustainable Development Strategy promotes sustainable development as a focused, broad, and a collaborative approach with some of the underlying core concepts as follows (Durgalakshmi et al., 2022):

1. Circumstances and all sides of life at all levels are affected by sustainable development.
2. The interests of humans including biological, material, and social interests, all are satisfied by sustainable development.
3. The factors responsible for damaging conditions of different forms of life are eliminated or notably reduced by sustainable development.
4. Sustainable development provides ambient solutions aiming to end hunger and poverty without comprising the future generation's need.
5. Resources are preserved by this approach.
6. Culture and the ecological systems are protected by sustainable development.

Sustainable development can be understood as the process of enhancing the standard of life of humans with the capacity to manage and contribute to the natural system. Sustainable development enables effective decision-making for the consumption and management of resources and the natural system. Sustainable development aims to achieve the UN's 2030 Agenda, but before stepping into the 2030 Agenda, there is a need to understand the Millennium Declaration, which was the primary plan for fighting poverty globally and the most adverse outcomes of the underlying concern (Fu, 2022). The agenda was approved in the year 2000, with the idea that Northern countries should contribute to the development of Southern countries through official development assistance (ODA). The aim followed by the Millennium Declaration was to achieve the gross domestic product to 0.7% of donors in order to lower the poverty in half by 2015 (Freeman, 2014). The Millennium Declaration has faced strong criticism and was unable to reach the underlying goals because of the association disclosure between the participating countries. Considering the situation, in 2015 the United Nations General Assembly officially embraced the document "Transforming Our World: The 2030 Agenda for Sustainable Development" (thereafter known as 2030 Agenda). Basically, the 2030 Agenda is guided by the 5 P's: people, planet, prosperity, peace, and partnerships (Figure 1.1) (Gordon, 2021).

- People: Eliminating poverty and hunger, irrespective of the forms, and guarantee dignity and equality.
- Plant: Protecting the natural system and resources of the planet as well as climatic conditions for the upcoming generation.
- Prosperity: Guaranteeing a flourishing and effective lifestyle with nature.
- Peace: Promoting a calm, peaceful, and comprehensive community or society.
- Partnership: Focusing on achieving the agenda goal by initiating global partnerships.

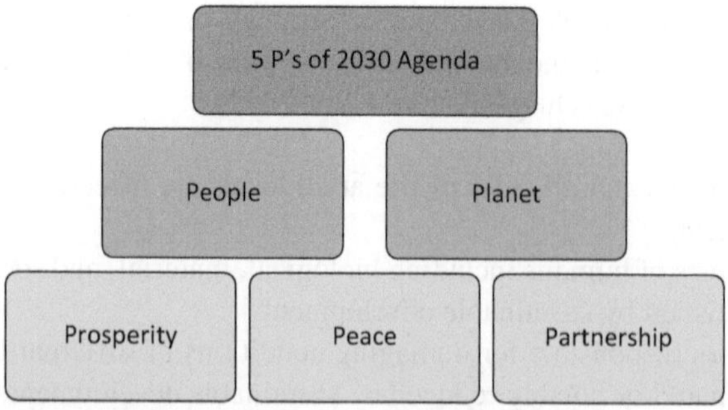

FIGURE 1.1 The 5 P's that should be considered in 2030 Agenda of Sustainable Development.

The most important primary objectives of achieving sustainable development are:

- To take care of the planet
- To ensure the well-being of society

There are seven basic priorities of the sustainable development plan when the needs and expectations of business stakeholders are met (Guandalini, 2020):

1. To fight against the changes in the climate and balance biodiversity
2. Look after the well-being of the society and people
3. Growth of a sustainable economy
4. Enforcement of digitalization and revolution for customers
5. Ensuring good guidance, transparency, and participation of stakeholders
6. Promote corporate social responsibility in the supply chain
7. Ensure good workplace environment and development of the individual

1.1.1 Definition

Sustainable development is development which meets the needs of the present without compromising the ability of future generations to meet their own needs (Hernandez, 2020).

According to the World Conservation Strategy report (1980), revealed by the International Union of Nature and Natural Resources (IUCN), for development to fall in the sustainable category it must take into consideration the social, economic, and ecological factors (Jones, 2021). The framework for sustainable development adopted today is strong, but still has some ways to go. The latest report of the Intergovernmental Panel on Climate Change (IPCC) states that major steps need to be taken immediately regarding the minimization of CO_2 emissions to maintain the temperature of the Earth below 2°C and thus preventing its damaging impacts. The major role in spreading awareness regarding sustainable development is played by UN, where various teams continuously work on a number of campaigns, such as #beatplasticpollution and #solvedifferent (Jones, 2021).

1.1.2 History of Sustainable Development

Sustainable development has become popular in the late stages of its evolution. It has become a useful guide to lead future trends. The following table illustrates the journey of sustainable development demonstrated in (Justice Mensah, 2019).

Year	Title	Description
1968	The tragedy of commons (Garret Hardin, 2010)	The ecologist and philosopher Garret Hardin, in his essay titled "The Tragedy of Common", argued that if individuals focused on their personal interests, this would deplete the planet's natural resources.
1972	Limits to growth (Dennis Meadows, 1972)	Dennis Meadows, commissioned by the Club of Rome, executed a computer simulation that targets to forecast the situation of what could be the state of the planet with limited natural resources. In the same year, the first UN Conference on the Human Environment was organized in Stockholm to discuss about the effect of humans on the environment.
1980	Human Development Index (HDI) (Renata Korsakiene, 2011)	Similar to the HDI, the combined solutions started after it was realized that the planet has limited resources and humans have to respect it.
1987	Brundtland Report (Gordon, 2021)	Also known as "Our Common Future", the Brundtland Report gave the remarkable and extensively acquired definition for sustainable development.
1988	Climate change (Rebecca R. Hernandez, 2020)	The International Panel on Climate Change was formulated with the aim to develop and grow awareness regarding the impact of humans on climate change.
1994	Triple bottom line (TBL) (John Elkington, 1994)	First used by John Elkington, founder of a sustainability consultancy firm, the assumption explains that companies should take into account three bottom lines in their businesses and should also consider the impact of their businesses on the planet.
2001	Millennium Ecosystem Assessment (Samper, 2009)	The investigation was initiated in 2001 by the UN and was four years long. The investigation reported and recorded that: • Humans have changed the ecosystem faster than ever before. • The changes to ecosystems have benefited humans individually and the economy, but adversely effected the planet and society. • Damages to the ecosystem will probably get worst during the 21st century.

1.1.3 Various Definitions of Sustainable Development

The two terms, *sustainable*, which implies "durability", and *development*, which aims at expanding the potentialities, together form *sustainable development* to address the problems in the environment that have occurred due to human activities (Juyal, 2016; Juyal, 2017; Juyal, 2020). By examining the definitions of sustainable development, experts have delineated the evolution of the concept. There have been many definitions proposed by different authors since 1987. Some of the definitions, presenting some common and different perspectives, are listed as follows:

- "Sustainable Development is meeting the needs of the present without compromising the ability of future generations to meet their needs" (Brundtland 1987).
- "Sustainable development involves devising a social and economic system, which ensures that these goals are sustained, i.e., that real income rises, that educational standards increase, that the health of the nation improves, and that the general quality of life is advanced" (Pearse, Makandia and Barbier, 1989).
- "Sustainable development is a system that can evolve indefinitely toward greater human utility, greater efficiency of resource use and a balance with the environment which is favourable to humans and most other species" (Harwood, 1990).
- "Sustainable development is a social construct, referring to the long-lasting evolution of a hugely complex system – the human population and economy embedded within the ecosystem and biogeochemical flow of the planet" "(Meadows, 1998)".
- "Sustainable development is a program for changing the process of economic development so that it ensures a basic quality of life for all people and at the same time protects the ecosystems and community systems that make life possible and worthwhile" (Van der Merwe, 1999).
- "Sustainable development is a form of economic growth which satisfies welfare needs of society in terms of short, medium and long term, it must meet the needs of the present without, however, compromising the future generations".(Viorel, 2002)
- "Sustainable development must be understood as a type of economic development that ensures meeting the needs of present generations without compromising the ability of future generations to meet their own requirements and applicable measures aimed at long intervals and long term effects". (Stefanescu, 2003)
- "Sustainable development is currently a powerful counter narrative to contemporary western lifestyles and forms of governing societies". (Beck & Wilms, 2004)
- "Sustainable development represents a shift in understanding of humanity's place on the planet, but it is open to interpretation of being anything from almost meaningless to of extreme importance to humanity". (Hopwood, 2003)
- "Sustainable development is a process of change, where resources are being gathered, an investment direction is chosen, the development technologies directed and various institutions have convergent actions, increasing the potential for human needs and desires". (Vare & Scott, 2007)

- "Sustainable development is seen as reconciliation between the economy and environment on a new path of development that would sustain the human progress not in a few places and for a few years, but on the entire planet and for a long future". (Sterling, 2018)
- "Sustainable development refers to the ability of a society, ecosystem, or any such existing system to operate continuously in an undefined future without reaching key resource depletion". (Marin, 2012)
- "Sustainable development can be defined as maintaining system stability by developing a balance of responsibilities: economic, social, environmental and technological support technique without compromising the needs of future generations". (Ivascu, 2013)

1.2 LITERATURE REVIEW

The approach of sustainable development has its foundation in 1968 when initiatives were begun in the underlying fields (Khan, 2021). Many studies have contributed to provide insight into the importance and need for the sustainable development, the goals associated with the concept, and how the holistic approach of sustainable development witnessed a suitable and adequate solution for the factors that result in the exploitation of the natural system and the natural resources, specially the scarce ones, and thus maintain a consistent ecological balance by fulfilling the demands of the present generation without compromising the needs of future generations. Various contributions were made in the field by maintaining the durable resources and the needs of humans by respecting the harmony of the environment (Kroll, 2019). David (2018) (Lam, 2020) highlighted the challenges in adopting sustainable development goals in India. The major concern is the lack of funding for achieving SDGs in developing countries. According to the report by World Bank (2013), per day 30% of its population was under the \$1.90 poverty. The gap in funding and the statistics associated with the SDGs is drawn from Figure 1.2 (Li et al., 2016).

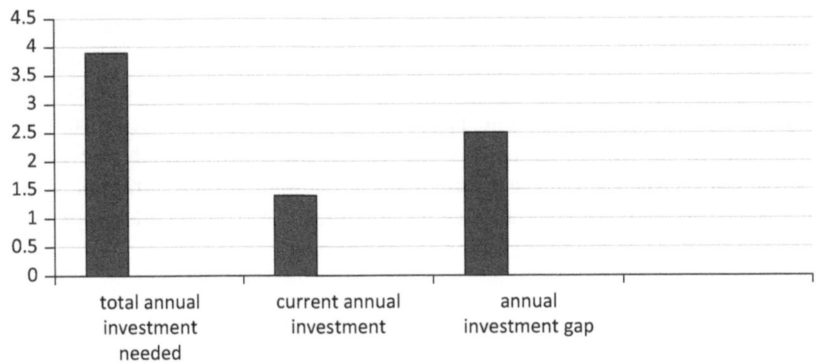

FIGURE 1.2 Gap in funding SDGs in developing countries.

FIGURE 1.3 The holistic impacts of digitalization and sustainability together.

The World Bank report (2013) states that by the year 2030 the cost in applying the SDGs in India is estimated to be US$14.4 billion and the annual funding shortage over 2015–2030 of about $2.5 trillion (Manikandan, 2022).

Mondejar et al. (2021) and David (2018) contribute to enabling a holistic approach for the digitalization benefits of sustainable development and address the serious challenges of plants that are endangered and also the change in climatic conditions (Figure 1.3). The holistic approach addresses the various aspects considered for the SDGs. The approach describes how digitalization will result in the changes for a more sustainable and balanced society (Maurya, 2022). Mondejar et al. (2021) examined the assistance of digitalization in adopting the SDGs in various sectors including the food–water–energy connection, industry, citizen's health and well-being, climate change, and biodiversity protection. Ozili (2022) illustrates the conceptual link between sustainability and sustainable development (also see Figure 1.4). The study demonstrates how development that is directed to the appropriate sustainability framework would result in sustainable development. The work refers to positive communication between sustainability and sustainable development. The framework explains that when sustainability is considered as the priority on the way to development, then sustainable development can be achieved. The sustainable development framework involves principles, guidelines, codes, laws, policies, regulations, and agreements. Any level of development that is achieved with the drawn framework is considered as sustainable.

1.3 COMPONENTS OF SUSTAINABLE DEVELOPMENT

For sustainable development, there should be the presence of sustained, organized, and coordinated efforts to result in socioeconomic development. There are three main pillars that form the underlying concept of sustainable development

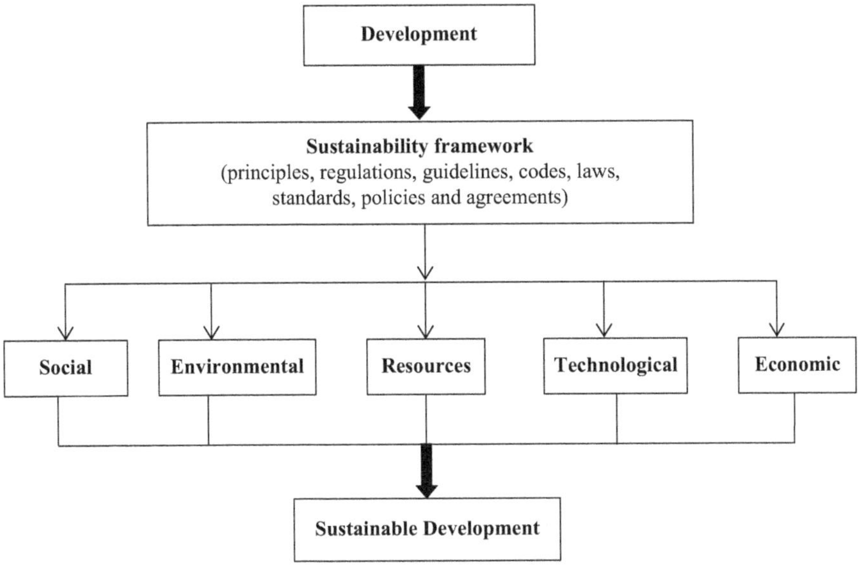

FIGURE 1.4 Conceptual relationship between sustainability and sustainable development.

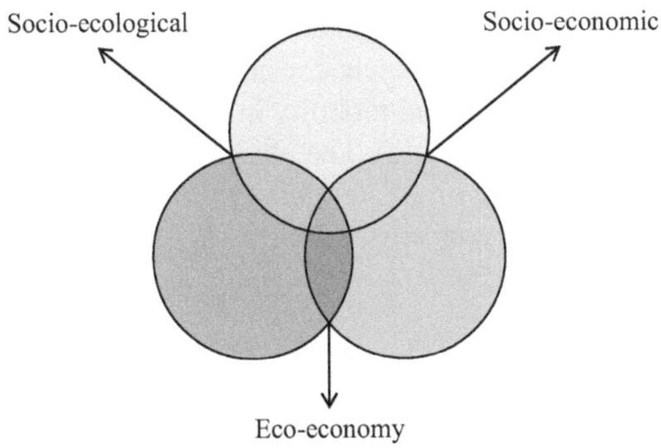

FIGURE 1.5 The three components contributing to sustainable development.

– economic, human (social), and ecological – built upon consistent growth of the economy, respect and advancements of social and human rights protection, and preservation of the ecology or ecological balance. All three components are interconnected and must be durable to make a sustainable world. The components and the subcomponents are illustrated in Figure 1.5.

1.3.1 Economic Sustainability

Economic sustainability implies the efficient and unbiased distribution of wealth. However, 40% of the world's population lives in poverty and it is almost impossible for those in poverty to earn enough to afford their minimum, basic needs. This has adverse effects on individual socioeconomic development as well as national development. Economic sustainability plays a vital role in healing these issues. The particular component addresses the maximum flow of income regarding the use of resources, especially limited resources. The economic component aims to make people think about building their economic well-being, but avoid doing any harm to the environment. In economic sustainability, all the decisions should be about long-term advantages rather than focusing on the short term. An important part of the economic component is strong financial support for:

- Universities
- Educational activities
- Research and development

The key points of economic sustainability are:

- Consistent and portable growth
- Risk management
- Total shareholder return

1.3.2 Social Sustainability

The concept behind the social component considers gender equality; development of people, communities, or culture; and mobility in order to maintain an unbiased and equal quality of life (Mohandas, 2018). The objective behind social sustainability is to provide and ensure the same or greater quality of life for current and future generations. Social sustainability concerns:

- Human rights
- Environmental laws
- Public assistance

The awareness and focus to this component have a great significance, as if we fail to take into account the same there will be a collapse of society. One initiative was the Water (Prevention and Control) Act, 1974, which organized India's central and state boards to examine water pollution and promote setting a minimum quality standard for drinking and surface water. The key points associated with social sustainability are:

- Dignity for the personal
- Equal opportunity
- Diversification
- Outreach programmes

1.3.3 Ecological Sustainability

Ecological sustainability takes into account the entire ecosystem, protection of the natural system and resources, and biological diversity. There are a number of situations occurring in the ecosystem that have an adverse effect on the environment, economy, and the society. So, in order to maintain ecological sustainability, these disturbances need to be avoided at every stage and wherever required. The ecological component involves proper balance and usage of natural resources, particularly of scarce resources. The points associated with this component are:

- Biodiversity administration
- Air emissions; use and discharging of water/chemicals

1.4 IMPLEMENTATION PROGRESS OF SUSTAINABLE DEVELOPMENT

In recent years, governments around the world have become more active in ensuring that sustainable development reaches every corner of the world. A global consensus was adopted to examine needs in order to analyze the efforts using the statistical capabilities in poorer countries. To empower the national statistical mechanism, in 2017, the Cape Town Action Plan for Sustainable Development Goals was produced. A joint survey was conducted in 2021 by the World Bank's Development Data Council and the United Nations Statistics Division (UNSD) and the Partnership in Statistics for Development in the 21st century (PARIS21). The strategic action adopted by Cape Town reported in the survey is illustrated in Figure 1.6.

The scheme proposed six major regions for the global action plan for sustainable development. The areas illuminated in the survey are:

1. Coordination and strategic initiative on data
2. Modernization of national statistical systems
3. Empowering primary statistical practices and programmes
4. Propagation and adoption of data
5. Multistakeholder coordination for data
6. Mobilization of resources and coordination of efforts for statistical capacity building

1.5 DIGITAL SUSTAINABILITY

The world has experienced digitalization in almost every sector. Digitalization has its importance in sustainable development as well. The concept and application of shifting the economic paradigm to a digital platform could be understood as maintaining and managing the economy in a long-lasting, organic way and preserving natural resources. The deployment of sustainable development digitally could lead

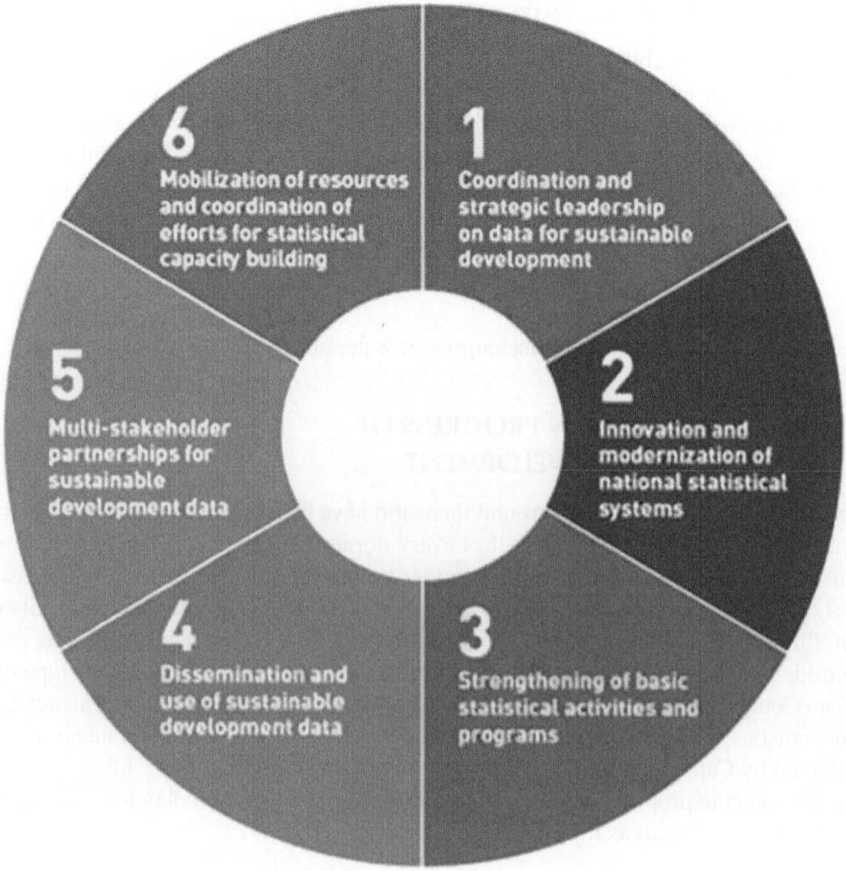

FIGURE 1.6 Strategic action areas of Cape Town Global Action Plan for sustainable development data.

to a swift and significant transformation in resource management. Sustainability has taken on notable importance and acceptance worldwide, and so for digitalization to take place, the integrative system enabling sustainable development digitally will have to have a remarkable future. The integration of digitalization and sustainability must develop a smart system of the Internet of Things to provide exclusive and strong opportunities to appropriately take into account the challenges underlined by the SDGs. Smart systems incorporating smart technologies are a game changer, especially for three broad elements: food, water, and energy. Also, digital technologies make it easy to tackle climate change. According to a report presented by Cognizant, in Southeast Asia alone, in the year 2020, around 40 million new users came online as compared to an average number of 20 million in the last five years. Every interaction done by individuals online, whether just a click, or pressing thumps up or down, has had significant effects on climatic changes. Video

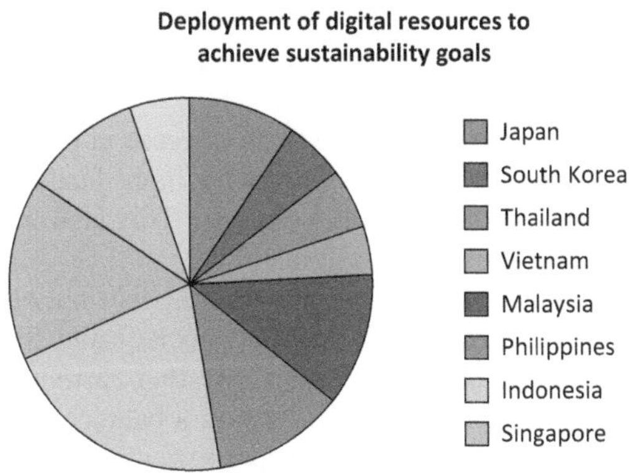

FIGURE 1.7 Statistics showing the deployment of digital resources across Asia Pacific countries.

streaming represent 4% of carbon emissions worldwide and is predicted to increase by 9% per year.

Digital sustainability is the practice of creating and using digital resources to maximize their value, keeping in mind the environment for current as well as future generations. TCS conducted a survey to analyze digital sustainability readiness among business leaders across Asia Pacific. The statistics in Figure 1.7 illustrate the deployment of digital resources across Asia Pacific in order to achieve sustainability goals.

1.6 CRITICAL FACTORS LEADING TO SUSTAINABLE INITIATIVES IN THE GLOBAL MARKET

With every passing year, industrialization and digitalization have achieved promising application around the world. The race for individual advantages and profits have adversely affected harmony with the environment, disturbing the ecological balance and causing the uneven distribution of resources. Some of the underlying factors for initiating sustainable development are:

- Global warming and climatic change: Rapid industrialization resulted in the degradation of forests, leading to the serious issue of global warming and adverse climatic changes. Thus, there is the need for sustainable development in order to build a green civilization.
- Overexploitation of natural resources: Human activities are the main cause for the uneven distribution of the natural resources, thus resulting in overexploitation of natural resources, particularly scarce resources. Awareness and knowledge about the controlled usage of limited natural resources needs to be shared worldwide.
- Poverty: Poverty is one of the most critical factors that enforces the development of the sustainable development globally. The unavailability of

resources, low income, and illiteracy contribute the most to the upsurge of poverty. Past efforts and progress to lessen the poverty rate have all been erased amid the COVID-19 pandemic.

- Degradation of cultivated land: Due to an increase in population and modernization of industries, there is a need for more land in order to build shelter, industries, academic institutions, etc. This land use has degraded cultivable lands, resulting in unsustainability.
- Ecological disbalance: The misuse and uneven distribution of natural and limited resources has adversely affected society and the environment. This use should be managed in such a way that current and future generations will both benefit. as well as ensure a balanced ecosystem. India is positioning itself as an economic superpower, but one has to consider some aspects of the country's economy. Figure 1.8 illustrates the profile of India. The illustration shows there is a need to address the importance of sustainable development. India has 17% of the world's population, but 36% lives below the poverty line and approximately 42% of the population is illiterate. Human activities continuously result in the depletion of natural resources, resulting in social, economic, and ecological disparities (Ahmad, Mishra, and Sharma 2023). The uneven distribution of resources, misuse of natural resources, adverse personal growth, and lack of harmony with the environment are among the number of reasons that cause the need for sustainable development. The major factors responsible for the need and importance of sustainable development are listed next and are shown in Figure 1.9.
- Climate change:
  - Increases in the temperature of the atmosphere and extreme weather events.
  - Growing sea levels due to the melting of glaciers.

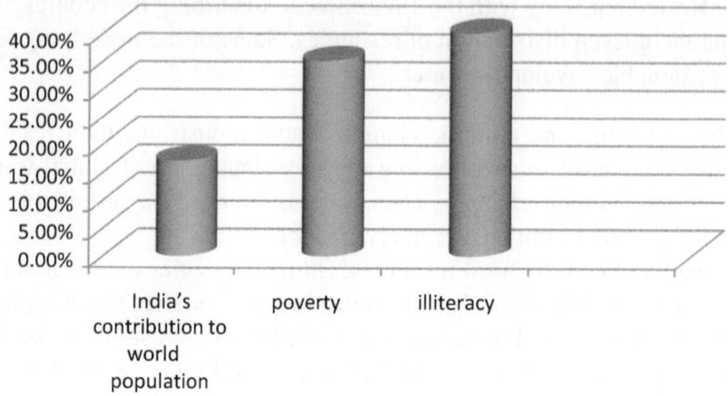

FIGURE 1.8 The profile of India that shows the distribution of various related issues in society.

FIGURE 1.9 The three major scenarios that create the need for sustainable development.

- Scarcity of resources
    - There is the need to double food production over next 38 years at a time when the count of the world's agriculture degraded land reach 23%.
    - By the year 2025, about two-thirds of the world's population will live in water-scarce/water-stressed areas.
- Overexploitation of natural resources.
- One million species are under the threat of extinction due to overexploitation of resources.

1.7 SUSTAINABLE DEVELOPMENT GOALS (SDGS)

According to the 2030 Agenda for Sustainable Development, 17 goals were embraced at the UN Sustainable Development Summit in New York in September 2015. There are 169 targets, 3506 events, 1326 publications, and 6499 actions associated with the SDGs.

The 17 SDGs are the global goals that are interconnected to each other and designed as a blueprint aiming to achieve a better and sustainable future for all. The SDGs provide a multidimensional view on the development. The SDGs are as follows:

Goal	Icon	Description
1		End poverty in all its forms everywhere. The pandemic period has erased all the progress over four years against poverty.

(*Continued*)

Goal	Icon	Description
2		End hunger, provide enhanced security of food and nutrition, and promote sustainable agriculture.
3		Ensure healthy lives and well-being for everyone at all ages.
4		Encourage equality in education, and ensure better and lifelong learning opportunities for all.
5		Attain gender equality and empower all women and girls.
6		Ensure availability and sustainable management of water and sanitation for all.

(*Continued*)

Goal	Icon	Description
7		Ensure economical, reliable, sustainable, and modern energy for all.
8	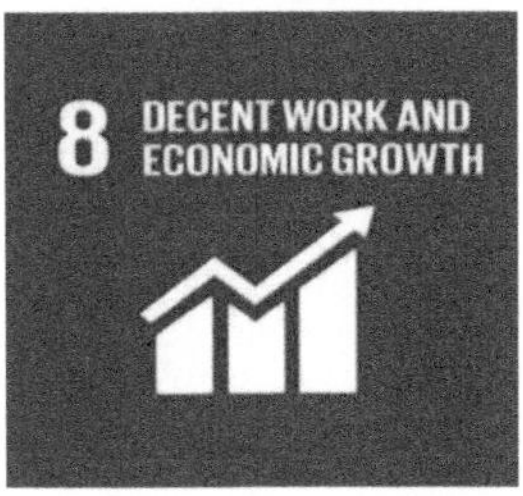	Foster comprehensive and feasible economic growth, productive employment, and decent work for all.
9		Construct strong infrastructure, promote inclusive and sustainable industrialization, and foster modernization.
10		Reduce inequality within and among countries.
11		Make cities and human settlements safe, strong, and sustainable.

(Continued)

Goal	Icon	Description
12		Establish sustainable patterns of consumption and production.
13		Take urgent action to fight against climate change and its influence.
14		Conserve and sustainably use ocean, sea, and marine resources.
15		Protect, restore, and promote sustainable use of terrestrial ecosystems, management of forests, and fight desertification.

(*Continued*)

Goal	Icon	Description
16		Encourage peaceful and inclusive societies for sustainable development, allow access to justice for all, and construct effective and accountable institutions at all levels.
17		Strengthen the execution and revitalize the global partnership for sustainable development.

1.8 SUSTAINABLE INITIATIVES IN THE IT SECTOR AROUND THE WORLD

With the Fourth Industrial Revolution, or Industry 4.0, there is a shifting from the traditional paradigm of pen and paper to digitalization. Everything in the information technology (IT) sector, whether sharing of resources, optimal use of resources, and so forth, have also taken a randomly different shift. Therefore, there is need to implement sustainable development for effective and seamless working of industries.

The implementation of sustainable development enhances decision-making. The most recent and pertinent explanation of sustainable development for industry is the triple bottom line (TBL). The concept of TBL works around three major foundations:

- Economic sustainability: The underlying objective is to prevent liquidity and guarantee profit.

Social sustainability: Shared access in the development of human and social capital.

- Environmental sustainability: It aims to the consumption of the resources that are generated by living and non-living things.

Sustainable development is considered as the incremental research on sustainable tools for fulfilling the needs and interest of the society at a high level.

There are other themes considered for sustainable development and Industry 4.0, as sustainable development requires a multifaceted approach. In addition to TBL,

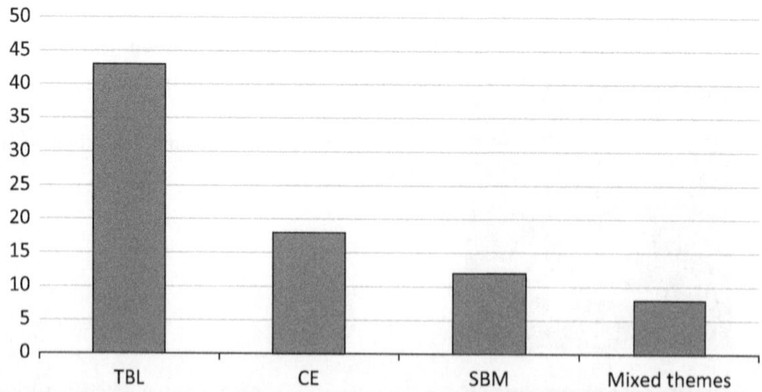

FIGURE 1.10 Key themes under sustainable development and Industry 4.0.

these key themes are a circular economy (CE) and sustainable business models (SBMs) (Figure 1.10).

There are other sustainable initiatives in rural and urban areas that are considered in terms of growing, replicating, transferring, spreading, and impacting higher institution levels. The initiatives carried out by the technology companies moves around six themes, which all together ensure sustainability and sustainable development in the IT sector. Figure 1.11 explains the themes and the initiatives of the technology companies. For example:

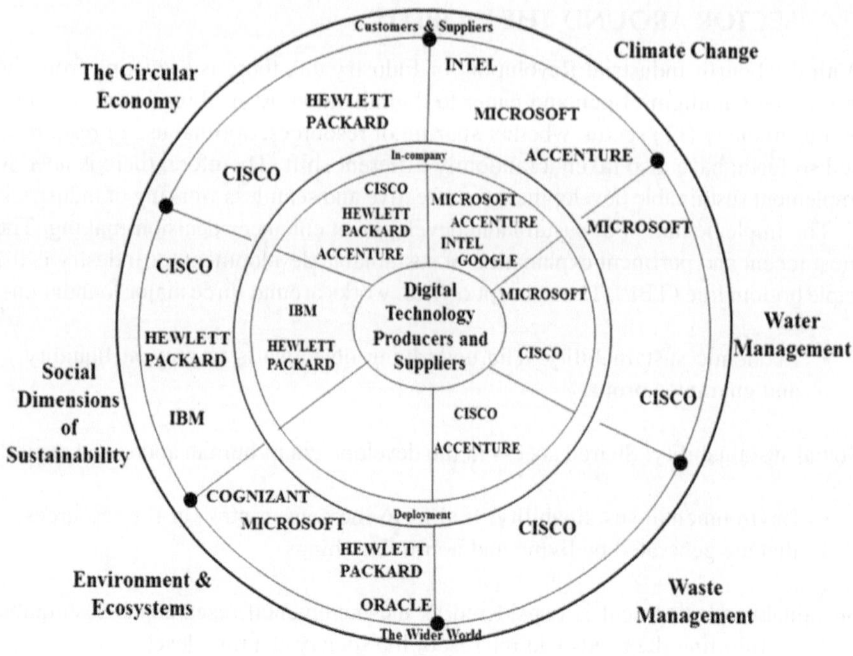

FIGURE 1.11 Themes of sustainable development in technology companies.

1. Accenture stated that climatic changes are a serious environmental and social concern. The company pledged to reduce the adverse effects on climatic conditions of its operations and products that emit carbon dioxide.
2. Google admitted that a huge amount of electricity usage is from its data centres, and hence is taking steps to transform its data centres into a more effective and energy-efficient systems by introducing smart temperature and lighting control, reconstructing the infrastructure to consume less power and energy, and enabling machine learning that will automatically optimize and initiate cooling of systems when needed. This initiative from the organization led to a sustainable IT infrastructure.
3. Another initiative headed by Microsoft Corporation, titled "Managing Our Water Use", concentrates on water resource management by reducing the usage of water in its operations, especially in water-stressed regions, and set a goal to make 1.5 million people "water positive" by the year 2030.

1.9 ROLE OF ARTIFICIAL INTELLIGENCE IN SUSTAINABLE DEVELOPMENT

Artificial intelligence (AI) was noted as the key facilitator for the aforementioned targets by meeting the needs of food, health, water, and services related to energy consumption. Enabling AI with other corelated technologies, smart cities would be an effective approach to reducing carbon emissions and meeting requirements of the power sector. Sixty-seven targets of the UN SDGs, 82% in the Society group, can be aided by artificial intelligence (AI) technologies to achieve the sustainable development goals (Maurya & Sharma, 2022; Juyal, Pandey, & Saggar, 2016, 2017; Juyal, Saggar, & Pandey, 2020). The specific sustainable goals where AI could be an effective are:

- SDG 1: No poverty
- SDG 4: Quality education
- SDG 6: Clean water and sanitation
- SDG 7: Affordable and clean energy
- SDG 11: Sustainable cities

The SDGs that would benefit from enhanced climatic conditions are:

- SDG 7: Affordable and clean energy
- SDG 11: Sustainable cities
- SDG 13: Climate action

However, there are 31 targets (38%) in the Society category that would be adversely affected by the influence of AI. But this is fewer than the targets with positive consequences.

The examination of the role of AI in sustainable development has a 79% positive impact and 35% negative impact (Ahmed et al., 2023; Durgalakshmi et al., 2022; Pushpalatha et al., 2022; Ali et al., 2022). The distribution of positive and negative

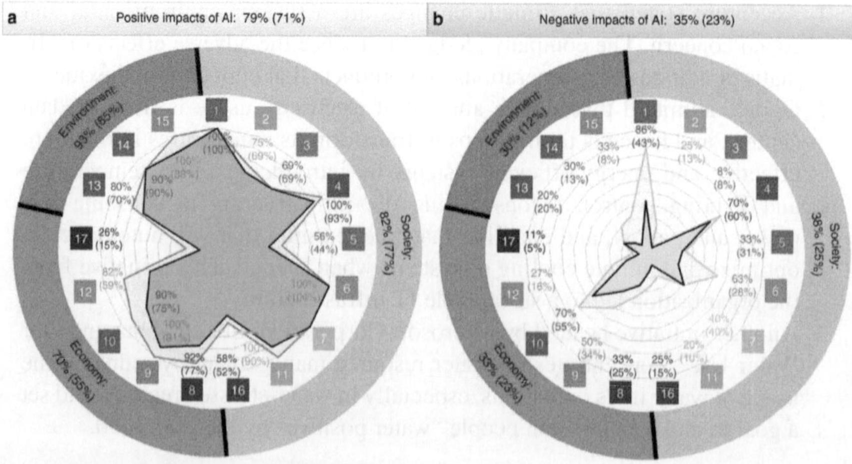

FIGURE 1.12 Illustration of positive and negative impacts of AI on sustainable development.

impact is associated around three broad categories: Environment, Society, and Economy (see Figure 1.12).

The positive impact of AI on sustainable development is as follows (also see Figure 1.13):

- The positive impact of AI on the environment is 90% (Singh et al., 2022).
- On society, the impact is 82% (Ali et al., 2022).
- On economy, it is recorded as 70% (Manikandan et al., 2022; Basheer et al., 2023; Sharma et al., 2023).

Figure 1.14 illustrates the negative impact of AI on sustainable development.

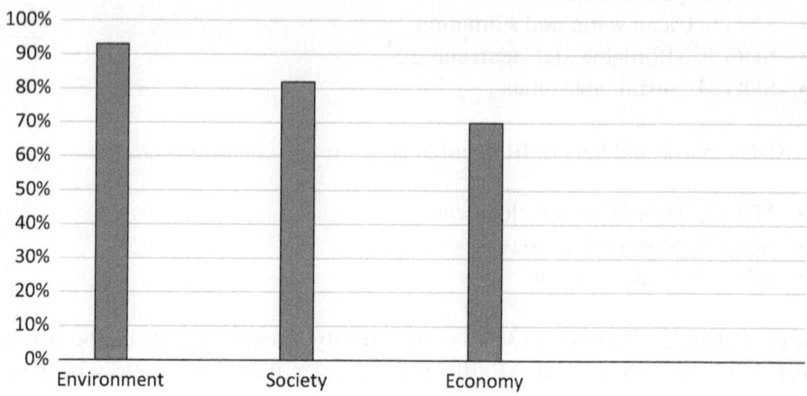

FIGURE 1.13 Statistics showing the distribution of positive impact of AI on sustainable development.

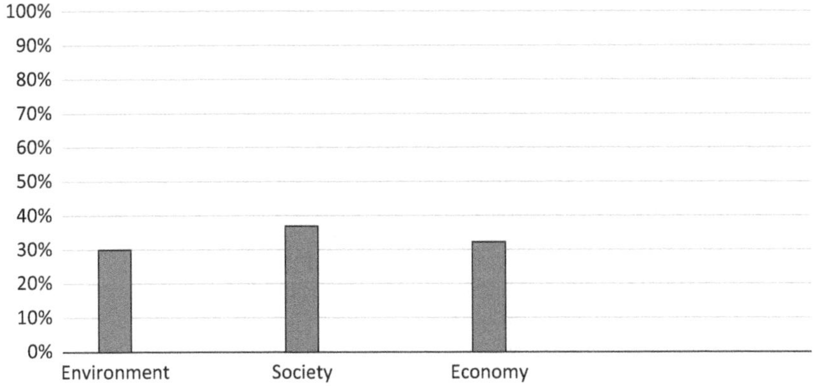

FIGURE 1.14 Statistics showing the distribution of negative impact of AI on sustainable development.

1.10 CONCLUSION

The population of the world is increasing significantly year by year, therefore the need for land, food, and resources is also gradually increasing, resulting in the degradation of the natural system and resources. Human activities are adversely affecting the ecological balance. Not only is life on land suffering from the unsustainability, but marine life is suffering as well. There is the need for sustainable development because scarce resources must be saved from overexploitation in order to meet the needs for future generations without compromising the present generation. People should raise their awareness to consider their personal economic benefit, while respecting and living in harmony with the environment. Industries and firms should go green in order to maintain the ecological balance, as the residuals from industries harm both human as well as marine lives. Various initiatives have be made for sustainable development and more are expected in the coming years.

BIBLIOGRAPHY

Ahmad, S., S. Mishra, & V. Sharma. (2023). Green computing for sustainable future technologies and its applications. In S. Grima, K. Sood, & E. Özen (Eds.) *Contemporary Studies of Risks in Emerging Technology, Part A (Emerald Studies in Finance, Insurance, and Risk Management)*, Emerald Publishing Limited, Bingley, pp. 241–256. https://doi.org/10.1108/978-1-80455-562-020231016

Ali, M. A., B. Balamurugan, R. K. Dhanaraj, & V. Sharma. (2022). IoT and blockchain based smart agriculture monitoring and intelligence security system. In *2022 3rd International Conference on Computation, Automation and Knowledge Management (ICCAKM)*, Dubai, United Arab Emirates, pp. 1–7. https://doi.org/10.1109/ICCAKM54721.2022.9990243.

Ali, M. A., B. Balamurugan, & V. Sharma. (2022). IoT and blockchain based intelligence security system for human detection using an improved ACO and heap algorithm. In *2022 2nd International Conference on Advance Computing and Innovative Technologies in*

Engineering (ICACITE), pp. 1792–1795. https://doi.org/10.1109/ICACITE53722.2022.9823827.

AO, A. M. (2023, May 23). *Intergovernmental Panel on Climate Change.* Australian Government, Department of climate change, Energy, the Environment and Water. Retrieved from DOCCEEW: https://www.dcceew.gov.au/climate-change/international-commitments/intergovernmental-panel

Bank, W. (2013). *The World Bank Annual Report 2013.* World Bank Group.

Basheer, S., K. U. Singh, V. Sharma, S. Bhatia, N. Pande, & A. Kumar. (2023). A robust NIfTI image authentication framework to ensure reliable and safe diagnosis. *Peer J Computer Science*, 9, e1323.

Bill Hopwood, M. M. (2003). Sustainable development: Mapping different approaches. *Sustainable Development*, 13(1), 38–52.

Constantí Stefanescu, J. P. (2003). Effects of climatic change on the phenology of butterflies in the northwest Mediterranean Basin. *Global Change Biology*, 9(10), 1–12.

David, M. P. (2018). Sustainable Development Goals (SDGs) – Challenges for India. *Indian Journal of Public Health Research and Development*, 41(2), 85–88.

Dennis Meadows, D. M. (1972). *The Limits to Growth.*

Digital Sustainability: How Digitally Green is Your Brand? (2021, July 22). Retrieved from Cognizant: https://www.cognizant.com/futureofwork/article/digital-sustainability-how-digitally-green-is-your-brand

Do You Know all 17 SDGs? (n.d.). Retrieved from United Nations: Department of Economic and Social Affairs: https://sdgs.un.org/goals

Duran, D. C., L. M. Gogan, A. Artene, & V. Duran. (2015). The components of sustainable development – A possible approach. *Procedia Economics and Finance*, 26(2015), 806–811.

Durgalakshmi, K. , P. Anbarasu, V. Karpagam, A. Venkatesh, B. Kannapiran, & V. Sharma. (2022). Utilization of reduced switch components with different topologies in multilevel inverter for renewable energy applications-A detailed review. In *2022 5th International Conference on Contemporary Computing and Informatics (IC3I)*, Uttar Pradesh, India, pp. 913–920. https://doi.org/10.1109/IC3I56241.2022.10073430.

Fu, I., & S. Schweinfest. (2022, February 3). *Survey on the Implementation of the Cape Town Global Action Plan for Sustainable Development Data.* Retrieved from unstatshub: https://covid-19- response.unstatshub.org/posts/survey-on-the-implementation-of-the-cape-town-global-action-plan-for- sustainable-development-data/

Elkington, J. (1994). Towards the Sustainable Corporation: Win-Win-Win Business Strategies for Sustainable Development. California Management Review. *Sage Journals*, 36(2), 90–100.

Elkington, J. (2015). Towards the sustainable corporation: Win-win-win business strategies for sustainable development. *Journal of Human Resource and Sustainability Studies*, 3(4), 90–100.

Freimann, A., M. Ham, & J. Mijoč. (2014). Measuring objective well-being and sustainable development management. *Journal of Knowledge Management, Economics and Information Technology*, 4(2), 1–8.

Gordon, J. (2021). *Brundtland Report - Explained.* The Business Professor.

Guandalini, I. (2020). Sustainability through digital transformation: A systematic literature review for research guidance. *Journal of Business Research*, 148, 456–471.

Harwood, R. R. (1990). The history of sustainable agriculture. In C. A. Edwards, R. Lal, R. M. Madden, & G. House (Eds.) *Sustainable Farming Systems*, Soil Water Conservation Society, Ankeny, IA, pp. 3–19.

Hardin, G. (2010). The tradegy of the commons. *American Association for the Advancement of Science is collaborating with JSTOR*, 162, 1243–1248.

Hernandez, Rebecca R., S. M. Jordaan, B. Kaldunski, & N. Kumar. (2020). Aligning climate change and sustainable development goals with an innovation systems roadmap for renewable power. *Frontiers in Sustainability*, 1, 583090.

Iglesias, E. (1987). *Report of the World Commission on Environment and Development: Our Common Future*. United Nations : NGO Committee on Education of the Conference of NGOs from United Nations.

Jones, P., & M. Wynn. (2021). The leading digital technology companies and their approach to sustainable development. *Sustainability*, 13(12), 6612.

Justice Mensah, S. R. (2019). Sustainable development: Meaning, history, principles, pillars, and implications for human action: Literature review. *Cogent Social Sciences*, 5, 1–21.

Juyal, Vandana, Nitin Pandey, & Ravish Saggar. (2016). A heuristic light weight security algorithm for resource constrained DTN routing. In *2016 IEEE International Conference on Computational Intelligence and Computing Research (ICCIC)*, IEEE.

Juyal, Vandana, Nitin Pandey, & Ravish Saggar. (2017). Performance comparison of DTN multicasting routing algorithms-opportunities and challenges. In *2017 International Conference on Intelligent Sustainable Systems (ICISS)*, IEEE.

Juyal, V., R. Saggar, & N. Pandey. (2020). An optimized trusted-cluster–based routing in disruption-tolerant network using experiential learning model. *International Journal of Communication Systems*, 33(1), e4196.

Khan, I. S., M. Ovais Ahmad, & J. Majava. (2021). Industry 4.0 and sustainable development: A systematic mapping of triple bottom line, Circular Economy and Sustainable Business Models perspectives. *Journal of Cleaner Production*, 297, 126655.

Kroll, C., A. Warchold, & P. Pradhan. (2019). Sustainable Development Goals (SDGs): Are we successful in turning trade- offs into synergies? *Humanities and Social Sciences Communications*.

Lam, D. P. M., B. Martín-López, A. Wiek, E. M. Bennett, N. Frantzeskaki, A. I. Horcea-Milcu, & D. J. Lang. (2020). Scaling the impact of sustainability initiatives: A typology of amplification processes. *Urban Transformations*, 3(2020), 1–24.

Li, L., Y. Lei, D. Pan, & C. Si. (2016). Research on sustainable development of resource-based cities based on the DEA approach: A case study of Jiaozuo, China. *Mathematical Problems in Engineering*, 2016, 1–10.

Lucian-Ionel Cioca, L. I. (2013). Sustainable Development and Technological Impact on CO2 Reducing Conditions in Romania. *Sustainability*, 7(2), 1637–1650.

Lucian-Ionel Cioca, L. I. (2015). Sustainable development and technological impact on CO2 reducing conditions in Romania. *Sustainability*, 7(2), 1637–1650.

Manikandan, N., D. Ruby, S. Murali, & V. Sharma. (2022, October). Performance analysis of DGA-driven botnets using artificial neural networks. In *2022 10th International Conference on Reliability, Infocom Technologies and Optimization (Trends and Future Directions)(ICRITO)*, pp. 1–6), IEEE.

Maria Rosario Perello-Marín, G. R.-G. (2012). Enhancing Education for Sustainable Development in Environmental University Programmes: A Co-Creation Approach. *Sustainability*, 10(1), 158.

Maurya, Aakash, & Vandana Sharma. (2022). Facial emotion recognition using Keras and CNN. In *2022 2nd International Conference on Advance Computing and Innovative Technologies in Engineering (ICACITE)*, IEEE.

Mazza, P. I. (2021). Concepts of sustainable development; a literature review and a systematic framework for connecting the role of education. *International Journal of Humanities Social Sciences and Education (IJHSSE)*, 11(21), 106–112.

Meadows, D. (1998). *Indicators and InformationSystems for Sustainable Development*. Hartland, VT, USA: The Sustainability Institute.

Meadows, D. (2015). Age-friendly cities performance assessment indicators system validation. *Psychology*, 6(5).

Meadows Donella, M. D. (2013). *The Limits to Growth*. New Haven: Yale University Press.

Merwe, A. S. (1999). The power of women as nurses in South Africa. *JAN*, 30(6), 1272–1279.

Mohandas, P. (2018). Sustainable Development Goals (SDGs) – challenges for India. *Indian Journal of Public Health Research and Development*. 9(3), 1–5

Mondejar, M. E., R. Avtar, H. L. B. Diaz, R. K. Dubey, J. Esteban, A. Gómez-Morales, B. Hallam, N. T. Mbungu, C. C. Okolo, K. Arun Prasad, Q. She, & S. Garcia-Segura. (2021). Digitalization to achieve sustainable development goals: Steps towards a Smart Green Planet. *Science of the Total Environment*, 794, 1–20.

Ozili, P. K. (2022). Sustainability and sustainable development research around the world. *SSRN*, 29.

Paul Vare, W. S. (2007). Learning for a change. *Journal of Education for Sustainable Development*, 1(2), 191–198.

Pushpalatha, N., S. Jabeera, N. Hemalatha, V. Sharma, B. Balusamy, & R. Yuvaraj. (2022). A Succinct summary of the solar MPPT utilizing a diverse optimizing compiler. In *2022 5th International Conference on Contemporary Computing and Informatics (IC3I)*, Uttar Pradesh, India, pp. 1177–1181. https://doi.org/10.1109/IC3I56241.2022.10072844.

Renata Korsakiene, I. B. (2011). Sustainable development and human development index. *Journal of Security and Sustainability Issues*, 1, 103–112.

Resources, I. U. (1980). *World Conservation Strategy: Living Resource Conservation for Sustainable Development. IUCN–UNEP–WWF.* The Environment & Society Portal.

Samper, C. (2009). The millennium ecosystem assessment: Science and policy for sustainable development. *Environmental Science*, 53, 1148–1149.

Sterling, S. (2018). Education for sustainability curriculum and pedagogy in higher education institution: Using complex adaptive system as a framework. *Creative Education*, 9(15), 49–70.

Sharma, V., B. Balusamy, J. J. Thomas, & L. G. Atlas (Eds.). (2023). *Data Fabric Architectures: Web-Driven Applications*. Walter de Gruyter GmbH & Co KG.

Singh, A., R. K. Dhanaraj, M. A. Ali, B. Balusamy, & V. Sharma. (2022). Blockchain technology in biometric database system. In *2022 3rd International Conference on Computation, Automation and Knowledge Management (ICCAKM)*, Dubai, United Arab Emirates, pp. 1–6. https://doi.org/10.1109/ICCAKM54721.2022.9990133.

Sustainable Development. (2020, January 20). Retrieved from Drishti IAS: https://www.drishtiias.com/to-the-points/paper3/sustainable-development-3

Sustainable Development – What Is It? Definition, History, Evolution, Importance and Examples. (2020, May 26). Retrieved from Youmatter: https://youmatter.world/en/definition/definitions-sustainable- development-sustainability

Viorel, L. S. (2002). Integrated Motor and Control Unit for Industrial Variable Speed Drive Systems. *Semantic Scholar*.

Zaras, K. M.-C.-T. (2012). Dominance rough set approach as a decision-making method for the selection of sustainable development projects. *American Journal of Operational Research*, 2(506), 1–13.

2 Patents and Innovations for Digital Sustainability

Sonu Kumar Jha, Udit Narayan, Rahul Kumar, Vandana Sharma, Naman Mishra, and Yogita Rawat

CONTENTS

2.1 Introduction ... 28
2.2 Need of Patents and Innovations for Digital Sustainability ... 28
2.3 Water Jacket for Solid Particle Solar Receiver ... 29
2.4 State of the Art of Autonomous Agricultural Off-Road Vehicles Driven by Renewable Energy Systems ... 31
2.5 Photovoltaic Technologies: Mapping from Patent Analysis ... 32
  2.5.1 Photovoltaic Cell Technology ... 33
  2.5.2 Patent Classification: Search and Organization Tool ... 33
2.6 Innovation in Sanitization ... 34
2.7 Optical and Thermal Performance Simulation of Micro-Mirror Solar Collector ... 36
2.8 Renewable Energy of Waste Heat Recovery System for Automobiles ... 38
  2.8.1 System Structure ... 38
  2.8.2 TEG Evaluation ... 39
  2.8.3 Effects of Temperature Contact ... 39
2.9 "Design and Development of Solar Umbrella Based on Peltier Effect" ... 40
2.10 Piezoelectric Based Shoe Charger ... 42
2.11 "Smart Traffic Roads Junction Energy Exploration for Electric Energy Generation" ... 44
2.12 How the U.S. Renewable Fuel Standard Could Use Garbage to Pay for Electric Vehicles ... 45
  2.12.1 Business Difficulties ... 46
  2.12.2 Technology ... 46
2.13 Conclusion ... 47
Acknowledgements ... 47
References ... 47

DOI: 10.1201/9781003348313-2

2.1 INTRODUCTION

In the context of digital sustainability, agriculture plays a vital role in feeding the population. The agrifood chain encompasses a wide range of duties, including soil tillage, planting, harvesting, storage, packaging, and procurement. However, the agricultural sector also has a significant environmental impact. Agriculture is responsible for feeding people. The industry performs a range of agrifood chain duties, from basic duties (such as soil tillage, planting, handling, cleaning, harvesting, drainage, and drying) to primary duties (such as storage and packaging) to tertiary duties (such as procurement). The Consultative Group on International Agricultural Research (CGIAR) reports that the agrifood channel alone absorbs nearly 30% of the world's total electricity generation and contributes about one-third (19%–29%) of global greenhouse gas emissions. Agricultural technology accounts for nearly 20% of global annual carbon emissions.

The World Energy Strategy sees minimizing environmental impact and reliance on fossil fuels as a key issue. Governments in many different countries are steadily encouraging the use of green vehicles including battery electric vehicles (BEVs), fuel cell vehicles (FCVs), plug-in hybrid electric vehicles (PHEVs), and hybrid electric vehicles (HEVs) [1].

Replacing vehicles with an internal combustion engine is in line with the 2050 greenhouse gas emissions (GHG) reduction targets set by the Intergovernmental Panel on Climate Change (IPCC). Using traditional fossil fuels to generate electricity has led to serious environmental problems such as acid rain, air quality, climate change, and global warming. But on the other hand, due to the increasing use of power tools and products in contemporary culture, the electricity consumed on the planet is also growing. According to this view, renewable technologies must be developed in order to compete pricewise with existing energy sources that cause environmental degradation.

Several countries have made the difficult decision to address these issues. The United States, for example, launched the Advanced Energy Initiative in 2006 with the ambitious goal of reducing imports and replacing 75% of its Middle East oil imports by 2025. This chapter revolves around these very concepts and covers the various patents which have changed the sustainable development, a few of them are explained below.

2.2 NEED OF PATENTS AND INNOVATIONS FOR DIGITAL SUSTAINABILITY

Most of us start by thinking about important issues related to sustainability, such as the importance of lessening carbon dioxide emissions, the transition to renewable energy, and the importance of eliminating wasteful items like polyethylene. These ideas become more common and better every time we think about how to solve environmental problems. In fact, there are several other options and strategies, including cycling and walking, converting to energy-efficient lighting, buying an electric or hybrid vehicle, or installing solar panels. Even though these should undoubtedly be the main priorities, it is crucial to consider how to make the internet and underlying communication networks more sustainable [2].

In fact, our increasing reliance on the digital environment and the continued expansion of bandwidth are two basic signs that the network is essential to everything we do. Networks and corresponding communication networks can help enable a more sustainable future, from basic communications to remote work and healthcare or online commerce [3]. Additionally, it is critical to consider the impact of people choosing to live and work online, and how we can make the system more sustainable in the face of growing demands for larger services and faster speeds [4]. The good news is that continuous technological development and new inventions that push the limits of physics make this possible, increasing the effectiveness of our networks and paving the way for environmentally friendly cyber strategies.

2.3 WATER JACKET FOR SOLID PARTICLE SOLAR RECEIVER

Solar energy is concentrated solar receivers for established solar concentrating methods using a field of heliostats (usually scaled on towers). The heat transfer medium passes via the solar receiver. The energy of the concentrated light is absorbed and heats this moving heat transfer medium. A continuous flow boiler receives the hot, flowing heat transfer medium to create steam that operates an electrical turbine generator and others.

A solar receiver module with a water jacket panel with a light-receiving side, a backside, and watered-enclosed plenum in the middle is shown in Figure 2.1. There

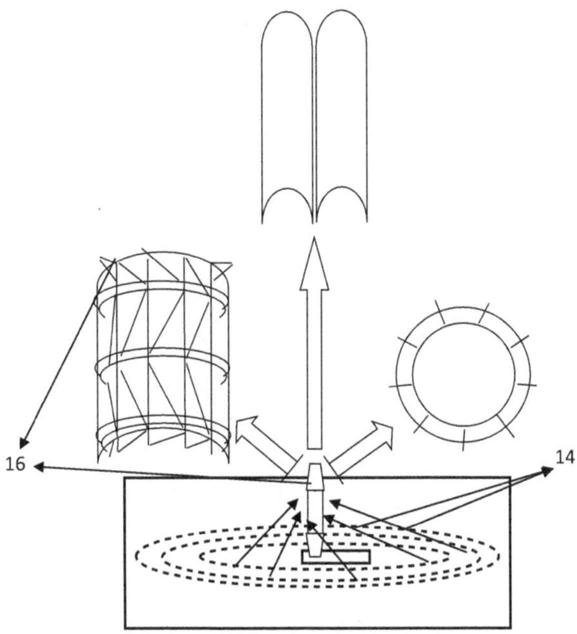

FIGURE 2.1 Water jacket for solid particle solar receiver.

is a light aperture going through the water jacket panel as well. The water-sealed plenum of the water jacket panel accepts light from the jacket panel's light-receiving side, as well as a light pipe that is optically coupled to the light aperture and extends away from the back of the water jacket panel.

The solar receiver includes a light channel tube having an aperture at the light-receiving end and extending into a volume through which a heat transfer medium; and a jacket containing an aperture; and a plenum through which cooling fluids other than heat transfer media flow [5].

The solar receiver of elements covered here is composed of a cylindrical assembly of panels, each of which has an aperture for light, a watertight plenum for the aperture, and an aperture per panel positioned to receive the light aperture. The cylindrical panel assembly of the plate features an outward-facing light-receiving surface [6].

The cylindrical rear panel is positioned inside the cylindrical panel assembly, forming an annular shape between the cylindrical panel assembly and the cylindrical back panel.

An annular gap between the cylindrical panel and cylindrical back plate with a light channel tube with an open first end and closed second end extends into the annular gap. The following provides more information on these nonlimiting features and the goals of the present disclosure.

A schematic representation of solar cell contractor system is displayed in Figure 2.1. An expanded detail view of the solar receiver is given in the inset and is denoted by a boxed arrow. The heliostats 10's multiaxis motorized system tracks the sun's path across the sky to collect solar radiation that strikes the heliostats and convert it into an energy beam 14 directed at the solar receiver 16. As seen in Figure 2.1, consists appropriate element includes beam-forming optics, which frequently contain mirrors or other reflectors, and beam-steering devices, such as multiaxis motorized systems, that work together to catch solar radiation impingement.

Figure 2.2 diagrammatically shows suitable welds used in the operation and an enlarged diagrammatic side sectional view of one light channelling tube of the assembled solar receiver module.

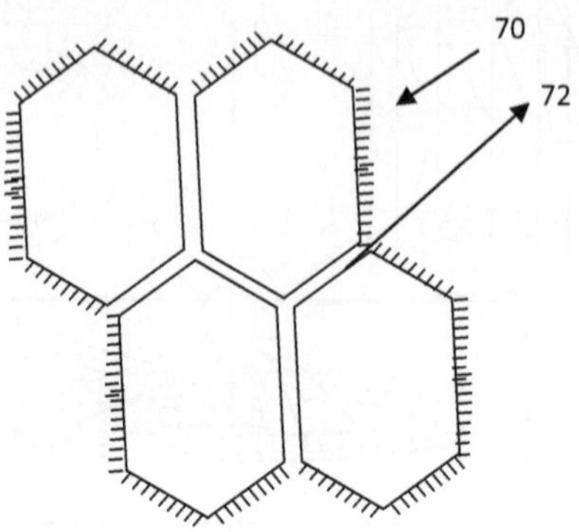

FIGURE 2.2 How the solar receiver module (from Figure 2.1) is formed.

2.4 STATE OF THE ART OF AUTONOMOUS AGRICULTURAL OFF-ROAD VEHICLES DRIVEN BY RENEWABLE ENERGY SYSTEMS

Agricultural complexes use electricity for various purposes, including product processing, crop handling, irrigation systems, farm buildings, and storage. Electric tractors are an example of agricultural machinery that uses electric propulsion. Agricultural systems can be used in machines and electrified to improve efficiency, adaptability, production, and performance [7].

One advantage is the ability to employ electronic auxiliary equipment, such as a radiator, cooling fans, air-conditioning, and water pumps to separate traditional mechanical driving components from the engine.

Many tractor manufacturers have recently upgraded their machinery, so that they can run on electricity and alternative fuel with biofuels, solar energy, and hydrogen fuel cells. Electric vehicles will thus be the main focus in this field in the future. Factors include high battery production costs, availability, and affordability. The ease of using fossil fuels hinder the wide use of electric vehicles. As a result, technological implementation and eco-friendly hybrid automobiles have been suggested as temporary fixes.

Unlike vehicles used in cities, agricultural vehicles are frequently located far from power sources and gas stations [8]. As a result, the cost of delivering electricity to these areas raises agricultural prices. However, autonomous on-site renewable energy supply systems can provide significant options while supplying farm energy demands or generating revenue from the sale of electricity to the local grid (Figure 2.3). As a result of dispersed generation and efficiency improvements, there is less reliance on fossil fuels [9]. The Ecosystem Integrity Fund (EIF) invests in start-ups that support environmental sustainability (Figure 2.4). The world's largest businesses are gradually improving their efforts in environmental sustainability. Over the next few decades, this trend will transform the economy.

The ability of the atmosphere to absorb heat or longwave radiation is caused by compounds of gases known as greenhouse gases (GHGs). The temperature of the surface of earth rises because of the presence of atmosphere which traps the escaping heat. Shortwaves or sunlight can easily pass through these gases and atmospheres.

Over the past 150 years, human activity has been the main driver of the increase in nearly all greenhouse gases in the atmosphere. In the United States, burning

FIGURE 2.3 Roadmap of on-site energy production.

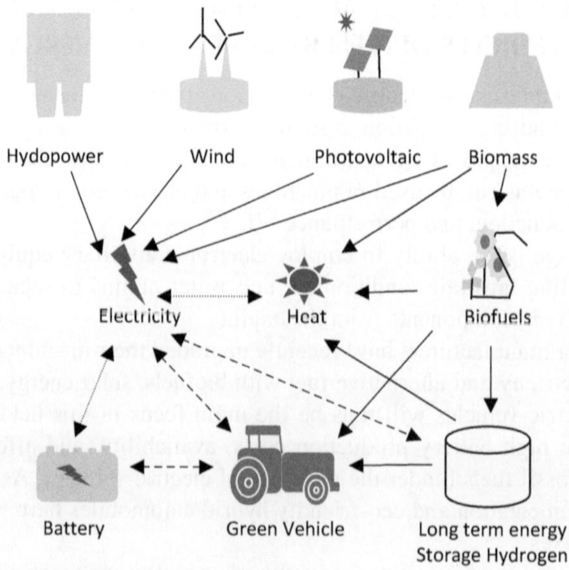

FIGURE 2.4 Schematic vision of EIF ecosystem.

fossil fuels for power generation, heating, and transportation is the leading cause of human-related greenhouse gas emissions.

2.5 PHOTOVOLTAIC TECHNOLOGIES: MAPPING FROM PATENT ANALYSIS

One reason not to use conventional fossil fuels for production of electricity is the resulting environmental problems such as air pollution, acid rain, global warming, and climate change. On the other hand, as more electricity-dependent items and technologies are used in modern life, more and more electricity is consumed globally. According to this view, the production of renewable energy systems is important to achieve affordable prices that compete with conventional energy origins that can damage the environment [10].

Some countries have worked hard to solve these problems. For example, the United States' Advanced Energy Initiative, launched in 2006, has a bold aim of diminishing its Middle East imports of oil in 2025 by 75% by developing alternative and sustainable energy.

Through analysis, this study aims to identify technological advances in solar cells. The survey covered 21,682 patents in the Derwent Innovation Index (DII) database. The research results show that (1) the number of photovoltaic cell patents is increasing yearly; (2) main depositors are the United States, Japan, China, South Korea, and Germany; (3) the United States and Japan have the largest numbers of registered patents; (4) the main scientific fields are engineering, instrumentation, energy and fuel, chemistry, and polymer science; (5) the semiconductor industry owns most of the photovoltaic cell patents; and (6) photovoltaic technology is based on polymers, nanostructures carbon, III-V compounds, cadmium telluride, and amorphous silicon cells. The use of solar plates for roof construction and generators to convert solar radiation into electricity is called photovoltaics.

Patent data is often used to evaluate and predict technology. Patents are rights granted to researchers, a means of technological development, and they reveal what a company believes is worth protecting. Part of the scientific and technical knowledge collected in a field can be represented by a collection of patents in that field. A good sign of technological progress is the increase in the number of technological patents. When concluding, it is important to carefully consider the total number of patents, as it may not be a reliable measure of the strength and ability of an invention.

2.5.1 Photovoltaic Cell Technology

Three generations are considered in the literature on photovoltaic solar cell technology. The first-generation systems (fully commercialized) use monocrystalline and polycrystalline silicon technologies. The second-generation depends on light voltage thin films and contains these entities: cadmium telluride, amorphous silicon, indium, copper indium selenide, and gallium Di selenide. The third-generation system includes natural photovoltaic technology that is still being demonstrated or yet not been widely sold, as the novel ideas. To include photovoltaic sources as renewable forms of energy, public awareness of the technological advantages (social, economic, and environmental) needs to be improved, as well as findings and innovation of new driven technology, and implementation of incentives for public photovoltaic policies and initiatives.

There should be more qualified workers available for the platform. Photovoltaic factory-improved terms for credible legislative frames are necessary to guarantee a turnover on investments, new innovation, and regular research.

2.5.2 Patent Classification: Search and Organization Tool

Recent research and its progress in certain scientific or technological sectors make use of patent information [11,12]. The Strasbourg Agreement of 1971 established classification as an integral part of patent applications. The World Intellectual Property Organization disseminates data on the world's patent distribution. Classifications are divided into sections, classes, subclasses, groups, and subgroups in a five-level structure (Figure 2.5). Each of the following groups are parted into major collections and subcollections, with the largest group accounting for 10% of the group, or about 7000. Given the goals of the study, it can be said that it is descriptive, which leads to a description of solar technology advancements. As for the methods, it is quantitative because variables must be quantified according to the set objectives to evaluate the data and report the results. In addition, improvement funds to encourage the use of sustainable resources in the home, part of that is the net measuring point. For example, California (US) uses internet measuring. China is developing

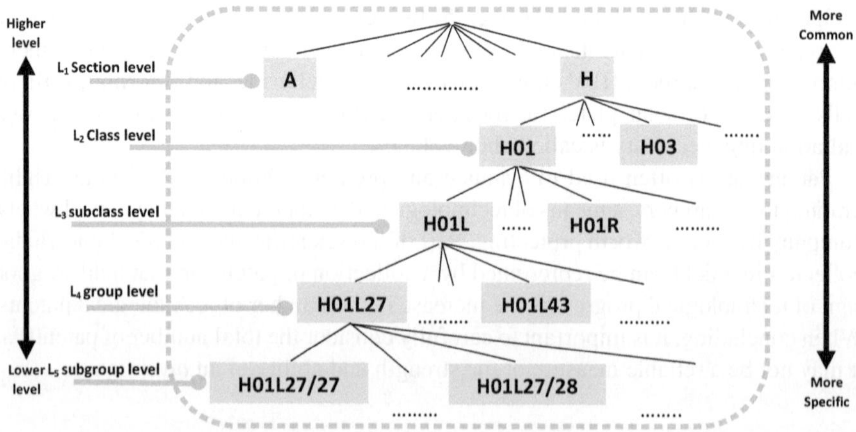

FIGURE 2.5 Hierarchical tree.

a township electrification plan to provide energy to rural areas. By using wind and solar photovoltaic technologies, the China Intellectual Property Office released the National Patent Development Strategy (2011–2020). Chinese regulations provide a range of incentives, including economic incentives, first-class research facilities, and tax incentives for companies that generate patents, while also focusing on individuals to encourage creative entrepreneurs in the field.

2.6 INNOVATION IN SANITIZATION

Figure 2.6 shows a flow diagram of a sanitization procedure. Depending on the liquid or chemical used to sanitize, the room may be locked to prevent someone from entering the area until the sanitization process and contact time are over. If the liquid is safe for humans, the door may not need to be locked. This process can be initiated in a variety of ways and by multiple parties, including through a building or property management system, the robotic device itself, an Internet of Things (IoT) hub (in conjunction with biosensor module 2406), or a remote user submitting a request through a networked computer device. Preplanned occasions, such as business meetings, school holiday parties, guest check-out times, and dinner seat reservations are other possible triggers [13].

Although a variety of powered disinfection and spray systems have been deployed, especially in healthcare facilities, these systems – such as stand-alone ultraviolet stations, commercial hand-held sprayers, or carts that require manual movement from room to room– are expensive or unavailable due to backlog. Additionally, because such portable devices require physical disinfection of a room or area, human error, duplication of tasks, or neglect of surfaces can occur. Large carts or treatment systems that remain stationary throughout treatment often require personnel to enter a room or area, putting them at risk of infection as equipment is physically moved from one location to another.

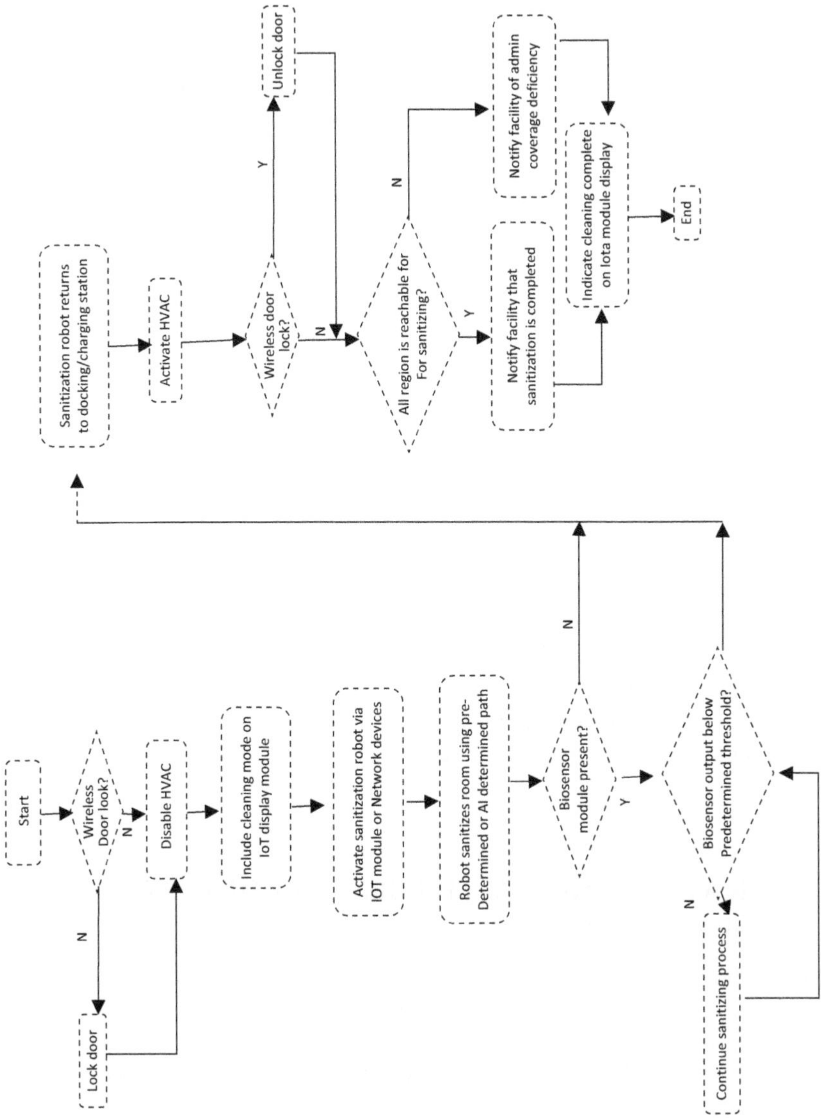

FIGURE 2.6 Flow diagrams of a sanitization procedure.

Most people either move from place to place because they can't adapt to changing conditions or rely on human power for cleaning tasks. The machine ideally sanitizes the area in a controlled and planned manner, minimizing human exposure to pathogen-infected rooms and areas compared to existing technologies. Early misting systems are ineffective because they cannot be used with air handlers, air conditioners, or other HVAC components that may affect airflow in the space being disinfected.

Current innovations are focused on methods and instruments to prevent and/or treat microorganisms and diseases in indoor environments, including COVID-19, suspected to be caused by the novel coronavirus SARS-CoV-2. The present invention more particularly relates to a device for disinfecting, sterilizing, or otherwise treating hospitals, hotels, classrooms, commercial spaces, and similar indoor spaces to reduce or eliminate viruses, bacteria, and other harmful pathogens, automatically reducing human contact.

It must be emphasized that the term *sterilization* as used herein refers to the reduction (completely or partially) of infectious agents or pathogens (e.g., viruses, bacteria, fungi, mould spores, helminths, and/or protozoa). The terms *cleaning*, *disinfection*, and *sterilization* have similar but significantly different meanings in the field. In general, cleaning refers to the act of removing impurities, such as dirt, bacteria, pathogens, and other contaminants from a surface, but not necessarily eradicating the pathogen in question. Disinfection is the use of chemicals, such as EPA-registered disinfectants, to remove bacteria from surfaces. Disinfection can help reduce the risk of infection after washing, but it doesn't necessarily "clean" dirty surfaces or remove germs.

Figure 2.6 illustrates robotic sterilization equipment, which follows specific paths throughout the cycle. In the architecture of a hotel there are several connecting rooms, walls, entrances, and furniture, in which the sanitization process occurs. The process is not just applicable to hotels but also to other facilities such as offices, hospitals, and educational institutions, with people using it during the day and night. With the process being completed, the robot then returns to its original location to charge.

2.7 OPTICAL AND THERMAL PERFORMANCE SIMULATION OF MICRO-MIRROR SOLAR COLLECTOR

Due to the world's growing energy requirements, various fossil fuels are used excessively, which worsens air pollution and contributes significantly to global warming. By 2025, the world's oil consumption is expected to surpass 120 million barrels per day, yet the supply of fossil fuels is expected to soon run out [14]. Consequently, switching to renewable energy sources may be the only action that is environmentally advantageous because they have less of an impact on the environment than other forms of energy.

Over the past few decades, efforts have been made to significantly reduce energy use. Around 35% of the world's energy is used in manufacturing. The construction sector has expressed interest in panels of solar temperature points. About 60%

of temperature resources required is used for an operation at a medium temperature (30°C to 250°C), comprising heating tasks like cooking, drying, and bleaching. Each country's share of the world's varying energy consumption is determined by the volume and nature of its industrial activities. Solar thermal systems are great alternatives to fossil fuels since they may provide reasonably priced household and industrial heating because to their high solar thermal conversion efficiencies. For this reason, during the past several years, the scientific community has continued to focus on creating such systems (Figure 2.7).

One of the criteria used to classify solar thermal systems is operating temperature. These systems are divided into three groups: low operating temperatures from 30°C to 150°C, medium operating temperatures from 150°C to 400°C, or more measuring thermal energy above 400°C. Depending on the track revolving line, solar plates temperature points are categorized as uniaxial, linear Fresnel reflector (LFR), or half circle, two-axis parabolic trough, parabolic collector, and also parabolic trough concentrator (PTC). Depending on the co-align solar plate light reflector, it is divided to centralized (PTC, LFR, etc.), decentralized (FPC, etc.), and evacuated tube collectors. Most concentrated solar collectors have a receiver, tracking mechanism, and concentrator. The receiver is positioned so that the condenser generates an image at the focus, focus axis, or plane of focal depending on the kind of condenser system being used. Examples of linear focusing concentrators include the following: (a) In order to aim the bunch of ray radiation in ratio of 60:1 to average incoming reflection, the LFR consists of a series of parallel mirrors, a receiver with a focus cavity shape, and one axis track point (from east to west). They can function between 100°C and 300°C. (b) Compound parabolic concentrators (CPCs) don't require tracking and can function between 100°C and 250°C. (c) PTCs need an east–west track point system and can function between 100°C and 400°C concentration ratios. A heliostat is a

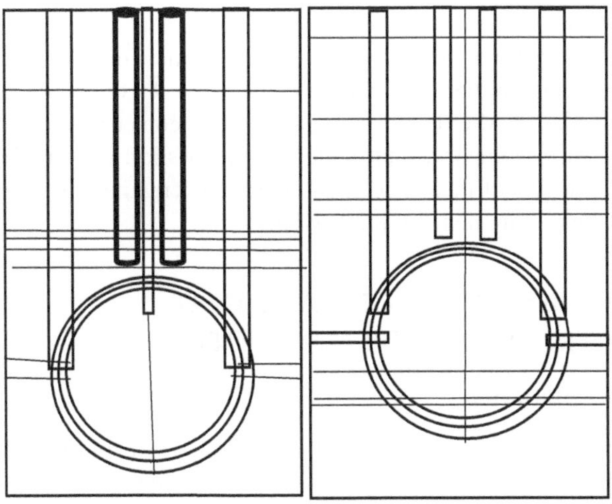

FIGURE 2.7 Models of absorber–tube connection.

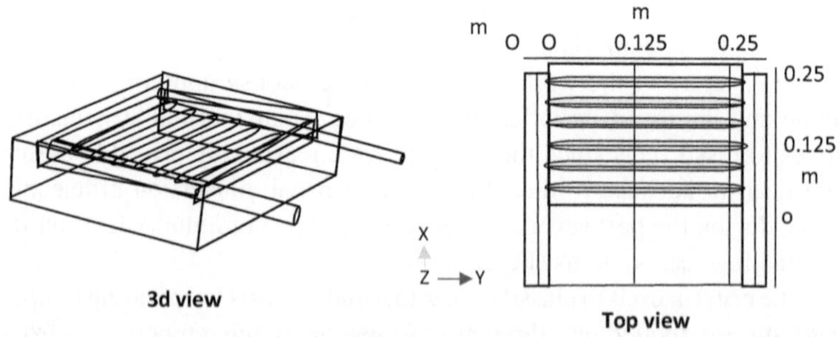

FIGURE 2.8 Receiver model in COMSOL.

huge array of two-axis surveillance mirrors that a central receiver uses to confine the beam's emission to a single point up to 1500 times.

This chapter's concentrator is a model for a special concentrating solar system that employs a static but internal monitored glass storage to direct incoming solar energy onto a plane type noticer. Panagopoulos's (2021) apparatus appears to be the most promising since it works with low to moderate thermal energy. Modelling is crucial for predicting its optical and thermal properties as well as for enhancing its long-term behaviour.

For ocular modelling and performance enhancement, geometric analysis techniques are most often utilized in the literature; however, some techniques also involve ray tracing. Though computationally rapid, simplified systems have the disadvantage of having a cap on the number of combinations that may be simulated. One of the various ray tracing methods that creates extraordinarily detailed simulations is Monte Carlo ray tracing (MCRT) procedures. The MCRT method is used to comprehensively and accurately model the optical performance of solar energy systems [15].

Two computer programs, Solstice and COMSOL Multiphysics simulation software, are used to examine the proposed system's beams. For calculations of thermal performance, we also utilize COMSOL (Figure 2.8). A command-line application called Solstice creates ASCII information and flux geometry can provide information about system's geometric and ocular characteristics. Ray tracing for various geometries and angles of incidence is automated by creating the necessary Python scripts to suit the requirements of this job [16].

2.8 RENEWABLE ENERGY OF WASTE HEAT RECOVERY SYSTEM FOR AUTOMOBILES

2.8.1 System Structure

To generate electricity that can be stored in batteries, we plan to build a device that converts the waste heat from a car's exhaust. This section discusses the modelling and implementation of thermoelectric generators (TEGs). To illustrate the idea

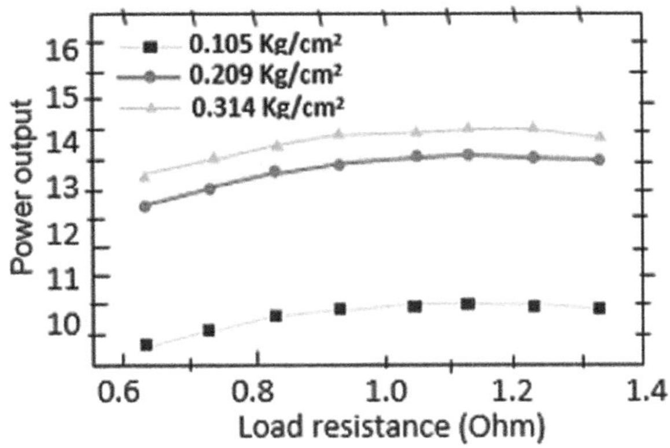

FIGURE 2.9 Load resistance and power input graph.

behind such a system, we tested it by creating a hot exhaust pipe model using the system's thermoelectric module. We use simulations to calculate the waste heat captured from the exhaust. The angular and travel belonging of each material includes an example are shown in Figure 2.9, which is helpful in understanding how waste is captured from the exhaust system [17].

2.8.2 TEG Evaluation

Prior to system construction, TEG evaluation is an essential stage in performance forecasting. In 1996, we examined the usage of TEG in excess heat cultivation using the idea of specific power. To demonstrate how power generation and temperature differential are related, Woo and Lee created a work simulation. In order to recover waste heat, our research aims to build a thermodynamic network of resistors, evaluate it, and use it. This is done with the aid of simulation software [18].

Through a series of simulations such as thermal field distribution, fin configuration design, and TEG module analysis, the results of simulation and experiment are verified and confirmed.

2.8.3 Effects of Temperature Contact

A more significant thermal-induced power can be produced, but improvement of the temperature and electric element degrades due to the thermal radiation depends on product belongings. The TEG measure points TGT8001K0SA0, Wise Life Technology, Taiwan, was created to compute the improvements of the TEG part. The TEG part is sandwiched between the top plate of copper and the heater's bottom. Two thermocouples are inserted to the plate of copper and the heater's top planes, respectively, and the thermal difference between the warm end and the 0 end is computed.

Figure 2.9 shows the power output with fixed temperature differential T = 110 or 303 kelvins on the coldest side with force per unit area 0.105, 0.209, and 0.314 kg/cm^2 and weight reactance 0.63–1.35.

Simulation: The four main parts of the waste recovery system are the hot side heat exchanger (HSHE), the cold side heat exchanger (CSHE), the thermoelectric module, and the zipped assembly points. To evaluate and develop the point, the CSHE fin arrangement and schematic are explained. In order to compare with the theoretical values, the simulation results of the power output of the TEG module under various temperature fluctuations and force per unit area of 0.105, 0.209, and 0.314 kg/cm^2 were observed. Figure 2.9 describes the dimensions and boundary conditions of each component in this series of simulations. The commercial computational fluid dynamics tool FLOTHERM supports the points improvements of different temperature settled options [19].

2.9 "DESIGN AND DEVELOPMENT OF SOLAR UMBRELLA BASED ON PELTIER EFFECT"

A series of flexible photovoltaic cells were installed on top of the study system, as its main goal was to build and operate a solar-powered thermoregulatory umbrella that could harness energy from sunlight hitting the tree canopy. Battery-based energy storage components should not be used in prototypes as they would make the prototype bulky and unwieldy. The solar energy from the photovoltaic cells will be processed by a DC–DC converter using a maximum power point tracking system, followed by a voltage regulation circuit to power the thermoelectric device [20].

A topology block diagram is shown in Figure 2.10. Its three main parts are the thermoelectric board, power conditioning circuit, and solar photovoltaic cell module

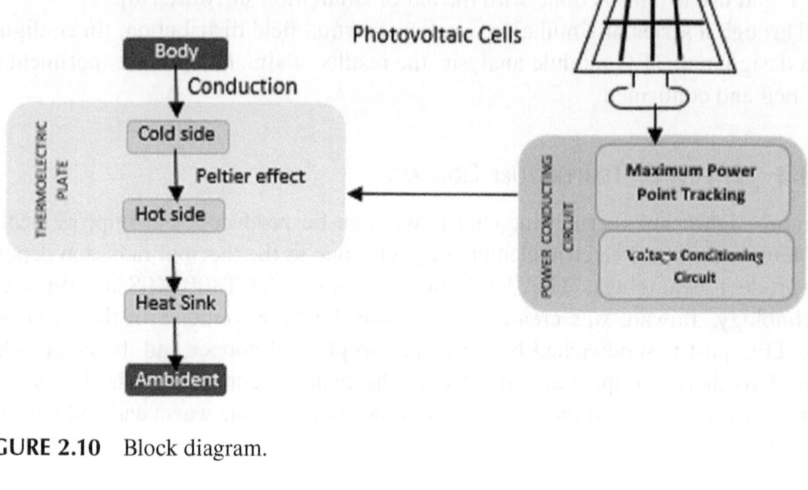

FIGURE 2.10 Block diagram.

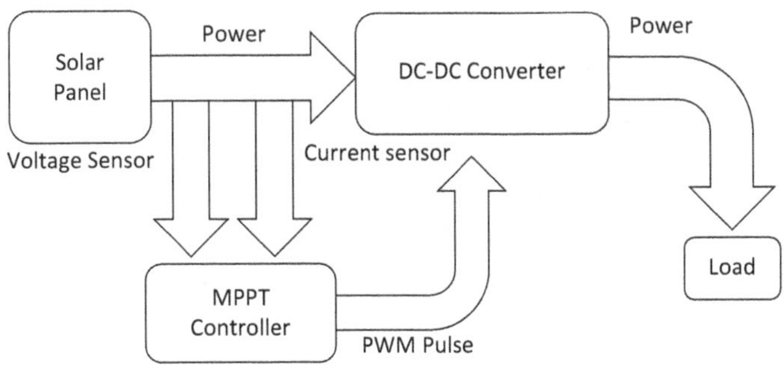

FIGURE 2.11 MPTT block based DC–DC converter block diagram [13].

[21]. Each of these components is described in later sections of the book. The fundamental building block of a system that harnesses the photovoltaic effect to generate electricity from light energy is the photovoltaic cell. It powers the internal electronics of the umbrella and acts as a source of energy for the batteries.

According to the temperature inside and outside, as shown in Figure 2.11, the voltage regulation circuit provides or modifies the power required by the Peltier or thermoelectric plate, which needs to be customized by the user.

The core components of the design are as follows:

The Peltier plate is installed between two metal tubes of rectangular cross-section along the central spine of the umbrella. One of the ducts will be used as an exhaust for the hot side of the Peltier plate and will be set up to release air from the space above the umbrella [21]. The second duct, which has a rearward curved outlet to allow air to be directed back to the user, is on the cold side and will serve as a cooling vent. Fans are installed on the upper ends of both ducts to provide constant air flow throughout the area [22].

Two thermoelectric plates are part of the design. Each thermoelectric module can consume approximately 6 watts of electricity. Therefore, the number of thermoelectric panels that can be used will depend on the intensity of the activity. Parallel electrical connections can be made between thermoelectric plates to improve cooling performance. Their final connection is through the control switch to the output of the voltage regulation circuit including the power driver. The control switch controls the current flow to the thermoelectric module (Peltier plate) of the circuit (Figure 2.12).

The research provides a mechanism for controlling temperature. By using thermoelectric cooler modules (Peltier plates) extreme temperatures can be recorded, controlled, and made available to users with temperature-resistant canopies. To control the heat released by this device as efficiently as possible, use a portable fan and heat sink. This technology replaces the methods used to handle heat dissipation in the past. Ultra-thin thermoelectric cooler modules offer a practical option as they are more competitively priced, require minimal effort, and take up less space.

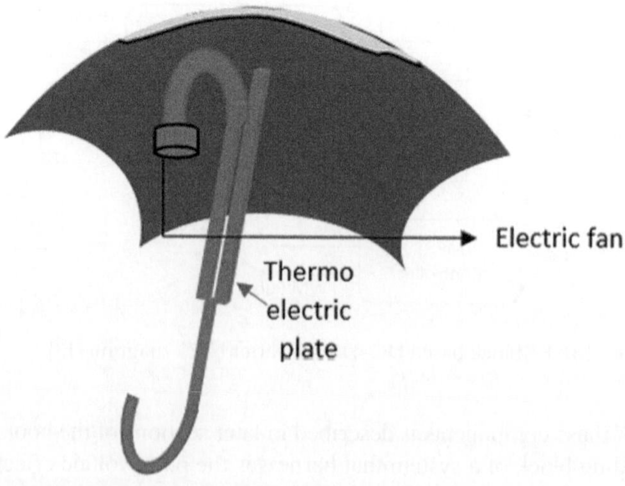

FIGURE 2.12 Proposed design.

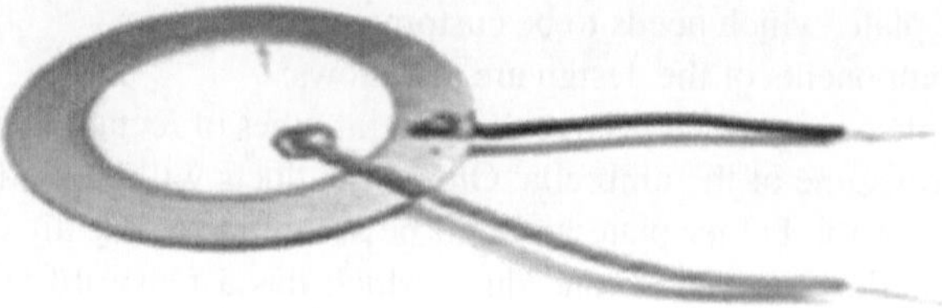

FIGURE 2.13 Electricity device.

2.10 PIEZOELECTRIC BASED SHOE CHARGER

A piezoelectric device [23] produces an electrical current from mechanical pressure using an electrical sensor (Figure 2.13). Instantaneous and opposing outcomes are two features of power materials. Other materials are subjected to mechanical stress, and as a result, an electrical charge is directly produced on their surface. On the other hand, some materials generate electrical charge without being subjected to mechanical stress. In the opposite consequence, pressure is applied to the associative object, which causes a charge to be created on the crystal's compressive face and an expanding face. The fabric becomes electrically charged when the pressure has been released [24].

FIGURE 2.14 Diode.

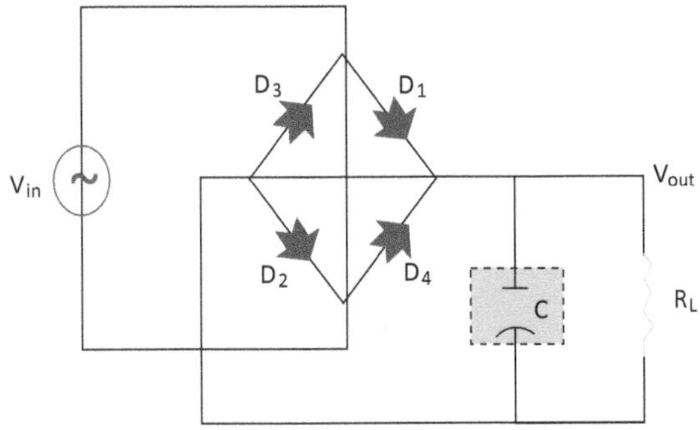

FIGURE 2.15 Circuit diagram of bridge rectifier (AC to DC).

A diode is a semiconductor that switches current in a single direction (Figure 2.14). It only significantly prevents current from flowing in one direction, while allowing it to flow in the opposite direction. As shown in Figure 2.15, we use diodes to build a bridge rectifier circuit in this project to convert alternating current (AC) into direct current (DC).

They are composed of substances such as crystals and ceramics, which are the only materials with the unique ability to convert mechanical energy into alternating current. Alternating current is a small disadvantage, though. Our project (DC) requires electricity. A full-bridge rectifier is regularly used as a rectifier part to convert the AC output of a power supply to a DC voltage. The rectifier circuit has four diodes. Since the supply must be continuous, the voltage must be rectified.

In contrast to liquid solutions, Li-ion reversible battery technologies (LiPo, Li-poly, Li-poly, LIP, etc.) can employ composite solutions. This mixture exhibits strong physical phenomena and is a solid (gel) polymer. A collection of cells has greater particular energy and thus can be used in implementations where weight may be crucial, such as smartphone electronics and heavier-than-wind aircraft. In this project the phones are charged with a power cord connected to the USB port

In this case, two piezoelectric sensor pairs have been connected. Each piezoelectric pair contains three sensors set linearly. Two sets of piezoelectric sensors are connected in a similar distant line with either to improve the output volt to 9V. The resulting line of the piezoelectric circuit supplies power to the bridge rectifier board [25]. A bridge rectifier transfers an AC source to a DC point. The voltage regulator controls the steady current required to charge the battery. The voltage regulator regulates the lesser 2.5 to 5 voltage current required to charge the battery. After the battery is charged, the phone can be charged using the USB port provided in the circuit where the charging cable is connected (Figure 2.16).

This way we can convert wasted energy in our daily routines into fresh energy and use it cleverly throughout the day. There is no additional cost or energy loss when using this energy. This could potentially transfer energy from human movement to objects. This electricity can be stored and used to power other systems.

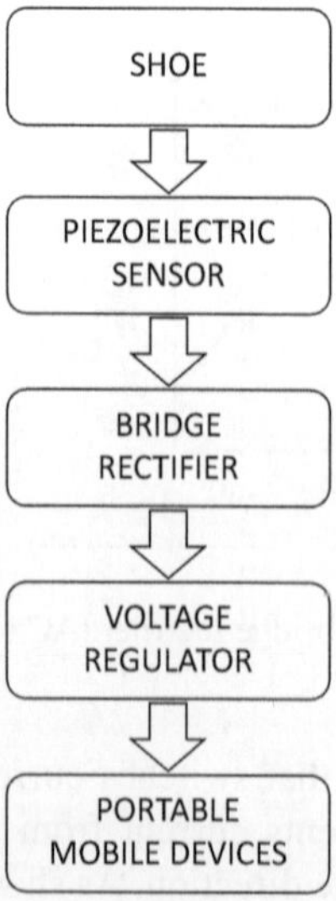

FIGURE 2.16 Block diagram of proposed work.

2.11 "SMART TRAFFIC ROADS JUNCTION ENERGY EXPLORATION FOR ELECTRIC ENERGY GENERATION" 27

The mechanical energy required for the movement of the transport system is generated through chemical energy [26]. There is no need to use the mechanical energy of the transportation system, including frictional, potential types of energy of the counterweight vehicle, thermal energy of tires in contact with the road surface, acoustic energy of the vehicle, and extra activities related to acoustic energy on the surface.

The piston spring, however, compresses the weight of the moving vehicle and converts it into useful electrical energy by using the proper procedures, such as placing piezoelectric transducers QB220-503YB (31.8 × 66.7 × 1.8 mm) on every particular lane of the junction. This place, known as an intelligent traffic junction, is where a number of significant road networks cross [3, 27]. Since traffic road crossings are simpler to overlook than single lanes, they were selected as the case study. It was discovered that the 2460 kg truck had the highest electrical production, therefore 100 transducers in four lanes would yield 102.612 kWh. The converted energy produced can power 1710 street lights or appliances within the allowed limitations given that LED street lights generally consume 60 watt-hours.

Most state capitals in Nigeria, including the nation's capital Abuja, frequently use some of their energy for ancillary uses including street lighting, vehicle traffic signals, and utilities like railroads and billboards. Maintaining these infrastructure results in an increase in the public load on the national grid, which results in higher utility costs for investors.

Roadways are a major civil framework that use much energy, about one-fifth of the overall energy used by the entire American civilization [28, 29]. Therefore, it is crucial to transition the transportation network to renewable energy sources as soon as possible.

When they are operating, the two main components of road traffic – the road surface and the automobiles – are exposed to a significant amount of renewable energy. Mechanical loads, solar radiation, and geothermal heat are three prominent renewable energy sources that are constantly present on a road surface. Energy origin provide a wonderful chance to obtain sustainable energy from sidewalks. Luckily, the improvement of energy harvesters offers a novel way to use the aforementioned potential ambient energy to generate electricity for the road traffic system. Furthermore, a number of researchers have verified their viability. In the 2000s, a present-day energy harvesting notion was put forward.

This is one of the procedures of capturing ambient energy and turning it to electrical energy. This includes the commonly wasted and dissipated solar, thermal, kinetic, and wind energy.

2.12 HOW THE U.S. RENEWABLE FUEL STANDARD COULD USE GARBAGE TO PAY FOR ELECTRIC VEHICLES 30

The US Renewable Fuel Standard is a substantial federal project that shifts the nation's carriage propellent mix toward low-carbon choices [30]. In 2014, a few types of renewable energy were added to the list of acceptable fuels under the standard, allowing the Renewable Fuel Standard (RFS) to support vehicle electrification. The largest origin of greenhouse gas (GHG) releases in the United States, cars created more than 28% of the country's carbon footprint in 2018. That percentage keeps increasing, which up to 2017, was the biggest contributor to greenhouse gas releases in the United States. However, the development of low-carbon power sources in the electricity power sector is making enormous strides.

Under the Energy Policy Act of 2005, the RFS is an important government programme designed to reduce the carbon intensity of the transportation sector by promoting renewable fuels. RFS results in the production of a specific amount of biofuel. For example, the Low-Carbon Gasoline Standard (LCFS) uses a technology-neutral framework and explicitly supports vehicle electrification to directly target the intended reduction in gasoline GHG emissions [29, 31].

The potential of the eRIN (electric renewable identification number) proposal is examined in this work. RFS makes money that will be used to significantly increase the use of carbon-negative and carbon-neutral energy from land gas filling, wastewater therapy facilities, and droppings. It will also be used to support the electrification of vehicles. Furthermore, it separates municipal solid waste [32].

Renewable Indentification Numbers (RINs) are being used to track and allocate renewable energy resources. In the case of ethanol, one RIN corresponds to one gallon. However, for electric energy, the current RFS plans have established that one RIN is equivalent to 22.6 kilowatt-hours (kWh) of electricity.

The following feedstocks are considered for RFS-compliant bioenergy supply: landfill gas (LFG), sludge from treatment of wastewater facilities (WWTP and MSW), manure, and municipal solid waste (MSW). We assume that the on-site

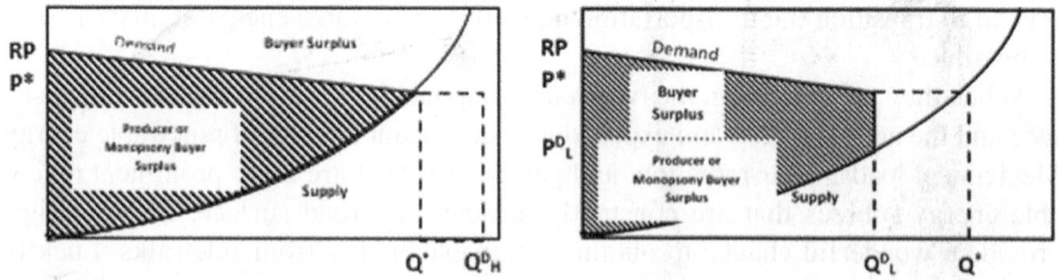

FIGURE 2.17 Supply demand and market characteristics.

biogas is fired by a cogeneration generator that also provides heat for anaerobic digestion (AD) procedures. Other agricultural wastes, such as corn stover that are anaerobic digestible, are not taken into account. Fats, oils, and greases (FOGs) are resources with enormous economic potential through codigestion [33].

2.12.1 Business Difficulties

As eRIN production and revenue grow despite the many advantages of our paradigm, there are some notable difficulties in designing and implementing programmes. First, the establishment of the biogas systems needed to sustain eRIN production by 2030 onward is aspiring. But, we found that when bioelectricity supplies fall, car subsidies might be reduced by only 2.8% and 16% in 2030 and 2035.

eRIN values may intentionally fluctuate from the levels predicted here. Our model expects the price of D3 (cellulose) RIN to continue at $2.50 as of now. The reduction in cellulosic fuel RFS will result in lower prices for D3 RINs and/or cellulosic exempt credits due to the addition of 21–240 million eRINs per year between 2025 and 2040.

This model may not be that large in practice, but it is a key component of the electric vehicle (EV) deployment trajectory. Instead of EV sales as a function of rebate magnitude, we model EV sales exogenously [34, 35]. However, in practice, the architecture will self-correct according to its market demand. Since rebates and sales growth are inversely related, if EV sales are lesser than expected, subsidies will increase, increasing the commercial appeal of EVs (Figure 2.17).

2.12.2 Technology

To evaluate project earnings and car subsidies, we simulated EV placement and bioenergy process curves, and use EV placement as an exogenous variable. The total project contains dimensions determined by dividing the overall fleet request (MWh) in year *n*–1 by the lower cost of bioelectricity supply provided at a rate lower the RIN value ($477/MWh) or the total fleet demand (MWh) in year *n*–1. Assuming that the remaining costs are borne by other links in the biopower-provided chain, including the indirect price of maintaining the chief accrual system, the cost of this biopower is deducted from the total project revenue, resulting in a 20% reduction.

2.13 CONCLUSION

The control versatility brought about by energy technology enhances the automation potential. The electrification of tractors and other agricultural equipment is critical in order to increase productivity, achieve energy autonomy, and ultimately reduce costs. Green electricity production in the field may accelerate the development of electric transport, thereby reducing greenhouse gas emissions. This report offers a patent evaluation of renewable technology advancement. The survey discovered that the top global depository nations are China, United States, Germany, Japan, and South Korea, and as such the patents for photovoltaic cells has been rising annually. Instrumentation, engineering, energy and fuels, chemistry, and polymer science are the primary fields of study, with the most significant universities and enterprises operating in Japan and United States. One of the industries that uses the most biofuels is agriculture, which not only protects it but puts its sustainability and yield at serious risk. Current trends in agricultural technology development do not meet optimal sustainable development of crop yields, creating new difficulties for the world's agrifood system. Likewise, revolutionary technological advancement movements to reduce environmental impact and improve systems seem essential.

ACKNOWLEDGEMENTS

Thanks to Galgotias University for providing the research environment.

REFERENCES

1. Ahmad, S., Mishra, S., & Sharma, V. (2023). Green computing for sustainable future technologies and its applications. In *Contemporary Studies of Risks in Emerging Technology, Part A* (pp. 241–256). Emerald Publishing Limited.
2. Juyal, V., Saggar, R., & Pandey, N. (2020). On exploiting dynamic trusted routing scheme in delay tolerant networks. *Wireless Personal Communications, 112*, 1705–1718.
3. Juyal, V., Pandey, N., & Saggar, R. (2017, December). Opportunistic message forwarding in self organized cluster based DTN. In *2017 International Conference on Infocom Technologies and Unmanned Systems (Trends and Future Directions) (ICTUS)* (pp. 497–502). IEEE.
4. Wasyluk, D. T. (2018, March 20). "Water jacket for solid particle solar receiver." U.S. Patent No. 9,920,955.
5. Torregrosa, A. J., Broatch, A., Olmeda, P., & Dreif, A. (2021). Assessment of the improvement of internal combustion engines cooling system using nanofluids and nanoencapsulated phase change materials. *International Journal of Engine Research*, 22(6), 1939–1957.
6. Ghobadpour, A., Boulon, L., Mousazadeh, H., Malvajerdi, A. S., & Rafiee, S. (2019). State of the art of autonomous agricultural off-road vehicles driven by renewable energy systems. *Energy Procedia*, 162, 4–13.
7. Scolaro, E., Alberti, L., & Barater, D. (2021, April). Electric drives for hybrid electric agricultural tractors. In *2021 IEEE Workshop on Electrical Machines Design, Control and Diagnosis (WEMDCD)* (pp. 331–336). IEEE.
8. Sacco, P., Gargano, E. R., Cornella, A., Don, D., & Mazzetto, F. (2021, November). Digital sustainability in smart agriculture. In *2021 IEEE International Workshop on Metrology for Agriculture and Forestry (MetroAgriFor)* (pp. 471–475). IEEE.

9. Pulido-Iparraguirre, D., Valenzuela, L., Serrano-Aguilera, J. J., & Fernández-García, A. (2019). Optimized design of a Linear Fresnel reflector for solar process heat applications. *Renewable Energy*, 131, 1089–1106.
10. Sampaio, P. G. V., González, M. O. A., de Vasconcelos, R. M., dos Santos, M. A. T., de Toledo, J. C., & Pereira, J. P. P. (2018). Photovoltaic technologies: Mapping from patent analysis. *Renewable and Sustainable Energy Reviews*, 93, 215–224.
11. Chen, Y. L., & Chiu, Y. T. (2013). Cross-language patent matching via an international patent classification-based concept bridge. *Journal of Information Science*, 39(6), 737–753.
12. Sampaio, P. G., González, M. O., de Vasconcelos, R. M., dos Santos, M. A., de Toledo, J. C., & Pereira, J. P. (2018, October 1). Photovoltaic technologies: Mapping from patent analysis. *Renewable and Sustainable Energy Reviews*, 93, 215–224.
13. Gorjian, S., Ebadi, H., Trommsdorff, M., Sharon, H., Demant, M., & Schindele, S. (2021). The advent of modern solar-powered electric agricultural machinery: A solution for sustainable farm operations. *Journal of Cleaner Production*, 292, 126030.
14. Panagopoulos, O., Argiriou, A. A., Dokouzis, A., Alexopoulos, S. O., & Göttsche, J. (2022). Optical and thermal performance simulation of a micro-mirror solar collector. *Energy Reports*, 8, 6624–6632.
15. Shubbak, M. H. (2019). Advances in solar photovoltaics: Technology review and patent trends. *Renewable and Sustainable Energy Reviews*, 115, 109383.
16. Sampaio, P. G. V., González, M. O. A., de Vasconcelos, R. M., dos Santos, M. A. T., de Toledo, J. C., & Pereira, J. P. P. (2018). Photovoltaic technologies: Mapping from patent analysis. *Renewable and Sustainable Energy Reviews*, 93, 215–224.
17. Kumar, A., & Rakshit, D. (2021). A critical review on waste heat recovery utilization with special focus on Organic Rankine Cycle applications. *Cleaner Engineering and Technology*, 5, 100292.
18. Tahami, S. A., Gholikhani, M., Nasouri, R., Dessouky, S., & Papagiannakis, A. T. (2019). Developing a new thermoelectric approach for energy harvesting from asphalt pavements. *Applied Energy*, 238, 786–795.
19. Desai, R. R., Chen, R. B., Hittinger, E., & Williams, E. (2019). Heterogeneity in economic and carbon benefits of electric technology vehicles in the US. *Environmental Science & Technology*, 54(2), 1136–1146.
20. Marati, N., Sandhya, C., Haarica, V., Gaggar, S., & Reddy, T. K. (2019). Design and development of solar umbrella based on Peltier effect. In *2019 International Conference on Computing, Power and Communication Technologies (GUCON)* (pp. 853–856). IEEE.
21. Fagehi, H., Attar, A., & Lee, H. (2018). Optimal design of an automotive exhaust thermoelectric generator. *Journal of Electronic Materials*, 47(7), 3983–3995.
22. Bharath, Y. K., Shruthi, K. H., & Dhavala, R. K. (2021, February). Peltier thermo-electric generator based standalone micro-power supply system. In *2021 7th International Conference on Electrical Energy Systems (ICEES)* (pp. 263–267). IEEE.
23. Khambalkar, S., Netke, S., Misal, T., Bhandakkar, P., Kohad, V., & Somkuwar, P. (2020). Piezo electric based shoe charger. *International Journal of Progressive Research in Science and Engineering*, 1(9), 8–11.
24. Ali, M. T., Khalid, T. J., Ahmed, A., & Nancy, M. R. (2019, November). Energy harvesting from human locomotion using piezoelectric transducer-based shoe. In *2019 IEEE international conference on power, electrical, and electronics and industrial applications (PEEIACON)* (pp. 39–42). IEEE.
25. Nimachwala, A., & Pandya, A. (2022). Energy harvesting in the form of electricity by Piezo–Cells by placing them in a pair of shoes and storing energy in Power Bank. *Materials Today: Proceedings*, *67*, 51–55.

26. Ibrahim, H., Ikotoni, B. F., & Jatto, I. Y. (2020, August). Smart Traffic Roads Junction Energy Exploration for Electrical Energy Generation. In *2020 IEEE PES/IAS PowerAfrica* (pp. 1–5). IEEE.
27. Juyal, V., Saggar, R., & Pandey, N. (2020). An optimized trusted-cluster–based routing in disruption-tolerant network using experiential learning model. *International Journal of Communication Systems*, 33(1), e4196.
28. Juyal, V., Pandey, N., & Saggar, R. (2017). Performance comparison of DTN multicasting routing algorithms-opportunities and challenges. In *2017 International Conference on Intelligent Sustainable Systems (ICISS)*. IEEE.
29. Juyal, V., Saggar, R., & Pandey, N. (2018). Clique-based socially-aware intelligent message forwarding scheme for delay tolerant network. *International Journal of Communication Networks and Distributed Systems*, 21(4), 547–559.
30. Younes, A., Fingerman, K. R., Barrientos, C., Carman, J., Johnson, K., & Wallach, E. S. (2022). How the U.S. Renewable Fuel Standard could use garbage to pay for electric vehicles. *Energy Policy*, 166, 112916.
31. Singh, A., Ali, M. A., Balamurugan, B., & Sharma, V. (2022, July). Blockchain: Tool for controlling ransomware through pre-encryption and post-encryption behavior. In *2022 Fifth International Conference on Computational Intelligence and Communication Technologies (CCICT)*. (pp. 584–589). IEEE.
32. Pushpalatha, N., Jabeera, S., Hemalatha, N., Sharma, V., Balusamy, B., & Yuvaraj, R. (2022, December). A Succinct Summary of the Solar MPPT Utilizing a Diverse Optimizing Compiler. In *2022 5th International Conference on Contemporary Computing and Informatics (IC3I)*. (pp. 1177–1181). IEEE.
33. Gorjian, S., Minaei, S., MalehMirchegini, L., Trommsdorff, M., & Shamshiri, R. R. (2020). Applications of solar PV systems in agricultural automation and robotics. In *Photovoltaic Solar Energy Conversion* (pp. 191–235). Academic Press.
34. Durgalakshmi, K., Anbarasu, P., Karpagam, V., Venkatesh, A., Kannapiran, B., & Sharma, V. (2022, December). Utilization Of Reduced Switch Components With Different Topologies In Multi-Level Inverter For Renewable Energy Applications-A Detailed Review. In *2022 5th International Conference on Contemporary Computing and Informatics (IC3I)*. (pp. 913–920). IEEE.
35. Sharma, V., Balusamy, B., Thomas, J. J., & Atlas, L. G. (Eds.). (2023). *Data Fabric Architectures: Web-Driven Applications*. Walter de Gruyter GmbH & Co KG.

3 Adoption of Sustainable Digital Technologies in Industry 4.0

Naman Mishra, Vandana Sharma, and Achyut Shankar

CONTENTS

3.1 Introduction 51
3.2 Artificial Intelligence 52
3.3 Cloud Computing 53
3.4 Big Data 55
3.5 The Internet of Things 57
3.6 Blockchain 58
3.7 Industry 4.0 in Different Sectors 59
3.8 Healthcare Industry 61
3.9 Industry 4.0 Challenges 62
3.10 Industry 4.0 Benefits 63
3.11 Conclusion 64
References 64

3.1 INTRODUCTION

Our current time is defined by the onus of technology in our everyday lives. Technology is, as per the *Cambridge Dictionary* definition, the practical implementation of scientific findings. The period we live in centres on the usage of stated scientific breakthroughs, spanning from but not limited to the internet, our mobile phones, computers, and the new age self-driven automobiles and so on. The implementation of technology in our daily lives have made our lives easier and more habitable, be it the usage of mobile phones, which have evolved from being just verbal communication devices to being used as devices to navigate as well as handle petabytes of data. We also use digital technologies for televisions, refrigerators and other home appliances, as well as self-driving cars. If we look upon our daily lives, we would see that nothing in our lives is untouched by technology.

If we now move on to new age technologies, we see that in our day-to-day lives we often hear words such as artificial intelligence, machine learning, and big data. We shall move forward by defining each of these unique technologies and how they

DOI: 10.1201/9781003348313-3

are shaping the world we live in. These words are not alien to our ears and we often use them in our conversations about our surroundings. *Artificial intelligence* is defined as "the capacity of a digital computer or a computer-controlled robot to accomplish activities that are generally associated with intelligent creatures". *Big data* can be summarized as "data which is humongous in its size, which can be moulded to understand the intricate relations between the different factors". *Cloud computing* can be described as "the sharing of resources over the internet by the use of well-established connections, which mitigates the use of physical accessories" [1]. These definitions give us a understanding of what we are dealing with in our daily lives as well as the factors that are heralding this new industrial revolution and ushering us into the era of Industry 4.0, whose bedrock comprises of these sophisticated and useful technologies. In a nutshell we define Industry 4.0 as the industry in which manufacturing, production, and distribution rely mainly on the usage of these new age technologies for better efficiency in terms of work and time management. To be able to further understand the use cases of these technologies and their contribution to the Industry 4.0 we would be looking at each of them individually.

3.2 ARTIFICIAL INTELLIGENCE

In layman's terms, artificial intelligence (AI) can be understood as using intelligence that is developed my machines and not humans, and letting the machines to do the thinking process as well. Over the years, artificial intelligence has been a hot bed of new research, which has helped this field to grow vastly and exponentially. This has allowed it to be the part of the intricate design of the World Wide Web, with applications in web search engines (for example, Google and Bing), the movie recommendation system of Netflix, speech recognition assistants such as Siri, and Tesla's self-driving cars. We see that these technologies are at our fingertips and are being used by the vast majority of the people at almost every given point of time. AI has found itself deeply embedded in our daily routines [2]. In terms of industry, we see that AI has already been helping us, for example, smart robots that adapt to environmental changes to balance power consumption as well as perform quality checks on behalf of humans.

Factories generate a lot of data, amounting to petabytes; this can be simply termed as big data. In the analysis of big data, AI plays a crucial role in that it is able to identify intricate relations and make decisions that increase productivity either by changing the course of production or making small changes that accumulate to greater benefits. For example, if a furnace that is controlled by AI can regulate its temperature based on the current weather, the company can save on energy costs. AI can also establish better relations in terms of data variables; if big data is fed to AI algorithms, they can figure out scenarios that promote customer relations and strengthen the customer base of a company.

One of the major applications of AI in industry is the automation of tasks, which reduces manual labour. With the help of AI, robots learn how to behave like humans in order to help us mitigate the amount of work that must be done manually. There is no need for a huge workforce on the assembly line, just a handful of people to look

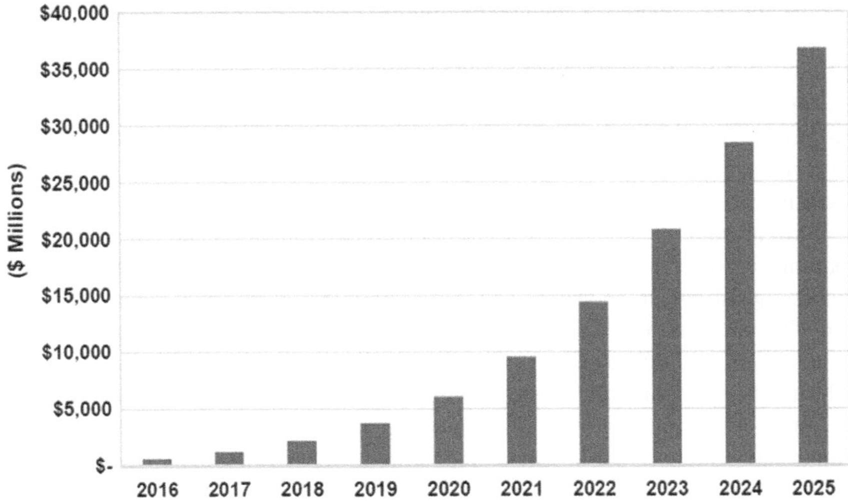

FIGURE 3.1 Revenue generated by AI over the years. From Tractica [4].

after the robots. Also humans do not need to adjust the machines, because this can be done by the robots as well [3]. In the field of healthcare, AI has application in finding the complex patterns in diseases, making it easier for healthcare professionals to make more accurate diagnoses and prescribe better treatment. Smart robots have started to take the place of humans in the operation theatre as well, with some major surgeries being performed by the intricate hands of robots. For example, IBM Watson analyzes patient data, then suggests a specially curated treatment plan. In the banking sector, AI is replacing human tellers and processing loan requests within seconds. In practice, these robots keep track of all our social activities and thereby suggest credible schemes that will benefit the company as well as the individual long term. Thus, we see how AI has become the nerve of modern industry and how it has potential to replace human manual labour.

Also note that the revenue generated by the contribution of AI was nearly $10 billion in the early 2020s, which is hefty in terms of a digital technology (Figure 3.1). However, the revenue generated by technology is expected to quadruple within a few years. This shows how big an impact AI has on the world in terms of economy showing its importance as well need for inclusivity in the industry. We also see that AI is the future of technology and it has already found uses in many sectors, and other sectors are thinking of ways to incorporate AI.

3.3 CLOUD COMPUTING

Cloud computing, in simple terms, is the use of hardware and software services through the internet, thereby eliminating the need of maintaining complex hardware in house [5]. The most common use of the cloud is for emails. We share and reshare thousands of emails over the internet with no storage being used up on our own personal computers. Cloud computing has paved the way for people to access complex software without the need for it being installed on the personal computer. It allows for real-time information exchange, without hampering the workflow. This real-time flow and sharing of information with people on the cloud sever has upped the game in terms of collaboration. Companies can collaborate to produce a particular product, enabling them to make smart decisions in real time as per the demands of the market [6].

If circumstances change or a problem emerges, companies can quickly respond, guaranteeing and modifying efficiency while lowering risk of repeating mistakes. Cloud computing can be termed as the driving factor of this new industrial revolution. The vast number of technologies that are dependent on cloud computing are countless and are different for different industries. For example, in the automobile industry, many sophisticated robots based on machine learning (ML) are controlled by cloud computing. When we talk about Internet of Things (IoT) technology, it has an intricate relationship with cloud computing by using it to modify the delivery as well as the analytical part of any industry. It has also helped by mitigating human interaction and ushering a new era of machines that are able to communicate with smart devices.

Cloud computing enables companies and organizations to adjust and grow more rapidly to new innovations, making them available to every opportunity that Industry 4.0 has [7]. Using cloud-based system enables organizations to have knowledge of modern information as well as the different manufacturing processes and apps ushered on by this new technology age without the need to establish costly, sophisticated infrastructure [8]. Cloud-based solutions may be readily altered and swiftly updated as company demands and technological offerings improve. Through the potential of cloud-based technology, companies may leverage the energy of technologically sophisticated apps and products and take their companies to greater heights.

Like earlier technological revolutions, Industry 4.0 seems to have the ability to dramatically modify and enhance the standard of living. But as with any new technology, it includes new situations and scenarios that will compel surrounding industry to change. Security difficulties are among of the most onerous issues confronting organizations today. For organizations shifting to the cloud, the necessity of competent, trustworthy security functions of cloud computing cannot be overlooked. Cybersecurity risks have always been present, and the common individual to huge companies are vulnerable. Hackers are shifting their focus to the cloud to compromise or alter systems, and do severe harm to companies [9]. Regardless of the sector or size, companies that are adopting cloud-based systems must keep in mind the various security flaws and hazards that are specific to the current era of cloud-based computing platforms.

The cloud promotes real-time interchange of information, generating and supporting an atmosphere of digital cooperation and integration. A company production line that employs the cloud and the real-time information provided by this system will be able to have better key players, offering real-time insight and enabling enterprises to consciously change and manage the production lines, improving efficiency and boosting risk management. IoT also helps by mitigating the complex interactions of different machines. The advent of IoT enables firms to gather and evaluate data via strategically placed sensors. The more meaningful data a company has access to, the more capacity it has to employ AI and build effective and scalable solutions.

3.4 BIG DATA

Big data analytics is the application of modern computer tools on massive data sets to identify important connections, trends, tendencies, and preferences for organizations to make more informed and productive choices. In the current digital era, big data insights act as a determining factor in every sector even in smart factories, where information collected from manufacturing equipment is examined to forecast when service and servicing procedures will be required. With the help of big data, companies and industries enjoy an efficient production line and productivity rate, combining this information collected with complex solutions, optimizing and automating the production cycle.

To put this in perspective, Figure 3.2 shows how much our data has expanded in terms of quantity. In 2014 when the whole collective data amounted to 5 zettabytes, it was considered a magnanimous task to understand the same. But over time, the amount of data has increased ninefold to about 45 zettabytes of data in the year 2020. To make it clear, 1 zettabyte of data is approximately equal to 10^{21} bytes of data, which is too big for the human mind to comprehend, thus the name big data. We also see that data is now a big part of our lives and every decision we make can be understood by carefully analyzing this huge amount of data.

Companies utilize big data to analyze the information in the same manner as most other business organizations albeit with a narrower emphasis [11]. They gather vast volumes of data from smart sensors using cloud-based sensors and IoT-based platforms, which enables companies to find trends, allowing them to enhance the effectiveness of supply chain management [12]. It also helps companies to find new and underlying factors that delay their production processes. When companies have

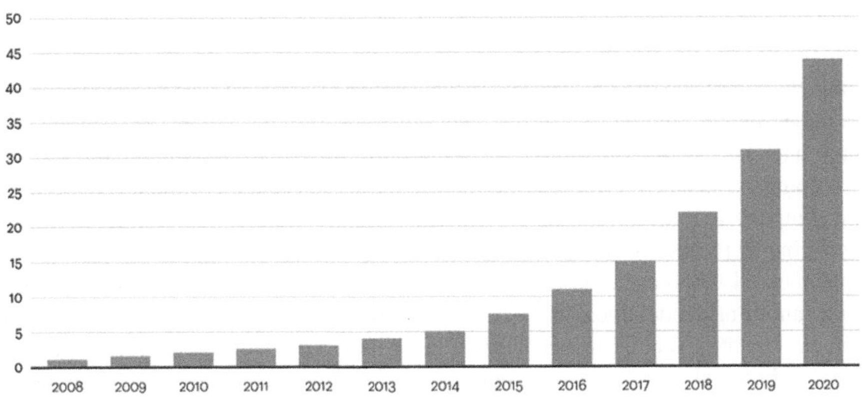

FIGURE 3.2 Volume of data in zettabytes [10].

identified these hindrances, then the companies can apply this acquired information to gain a better insight to these hidden reasons. This in turn allows firms to improve productivity while cutting costs and minimizing waste [13]. Production efficiency and resources are vital to companies. Sustaining their productive resources and keeping schedules tight and on track might make the difference between a stellar reputation and a negative one.

Big data analytics decrease malfunctions and unexpected downtime by roughly 25%. Using self-service analytics in engineering may condense vast bulks of big data from industrial facilities. For example, the multinational chip-maker Intel has been using big data to get the necessary real-time data to analyze and further its manufacturing processes. The smart system then segregates the data, applying the useful data, and discovers trends, diagnoses errors, and generates visuals for important decision makers. Engineers employ big data analytics output produced from the system to make judgements. With this information, they prioritize modifications and steps to be performed to minimize unexpected downtime or equipment breakdown.

Industry manufacturers utilize big data to manage and ramp up their production processes. The first step in this process is to mitigate the work done by humans and lessen human involvement required in a production process. This analysis mainly works by analyzing the data present in the database, the flaws of the process, and the limitations encountered, connecting the dots and the findings with what is currently going on inside the factory. These findings guide the automated robots to take the place of humans, limiting the number of errors as well as guiding them to the perfect strategy to maximize production. We see that when the data has been meaningfully actuated and the results that are derived are used to guide these robots, these robots are the ones that now handle the production process making the process error free as well as tweaking the process when any new information is discovered, keeping the process efficient.

A very high and rising proportion of business operations happen via this new era of internet communication. With this kind of integration, the amount of data collected becomes humongous, inclusive of both what is on the internet and off the internet, which makes the process very complex as well as impossible for enterprises to keep track of the data and make judicious use of it. When we look back years ago, industries were confined to small regions and catered to the needs of the local public and the local market. But times have changed drastically and today there are no geographical boundaries. In this new competitive environment, knowledge and data, and the judicious use of them is the only thing that keeps companies alive. If a company decides not to use big data, then its existence is threatened because the company will lag behind its competition who are making new market discoveries owing to the use of big data [14]. We also see that anything new should not be implemented without the proper planning, otherwise resources are wasted and not doing anything productive. On the other hand, if a company is able to understand the potential of big data and incorporates it in its daily proceedings, then the company stands a better chance to show its potential and compete in this new era of industry. Big data helps to ramp up the production process by linking the different sectors, finding the relations between the sectors and combining them to boost the process.

When big data is employed, there is automation in collection of data, allowing for processes to become more and more refined with the help of the analysis provided.

3.5 THE INTERNET OF THINGS

To define Internet of Things, we must start with understanding the "things". These things are the physical hardware systems that are kept in the loop and connected with the help of sensors. These sensors store data and communicate with the other devices based on the data collected with the help of the network [15]. This establishes a well-connected and maintained ring of hardware that works in collaboration with one another without the need of human interaction. These gadgets vary from commonplace home products to complex industrial machines. IoT may be described as a network whereby all the electronic devices that we use can communicate with one another and transmit data utilizing the network without any human intervention. The items are often integrated with sensors and thus connect with the other gadgets through the internet. The IoT takes the things we use daily a step further by connecting them to the World Wide Web, while incorporating additional processing capabilities and using the data to learn meaningful things about the surroundings [16]. In simpler terms we can say IoT is a web of linked devices. These devices capture and communicate data pertaining to their environment in which these devices are deployed and used. A significant element of IoT is the large amount of data that it creates or draws. Appropriate data management is consequently a key component to determine if the vision of IoT becomes a reality.

With the advent of current technology, gadgets and intelligent systems that are shared over the network will usher a new era for the industry as well as our daily lives [17]. Future IoT will encompass an array of sophisticated sensors that can contemplate new risks in the industry, such as the different hazards of a warehouse [18]. The network of suppliers is also managed by IoT to ease the insurance process and gauge the risks involved for managers and officials, mitigating the involvement of attorneys [19].

Consider the situation of automotive insurance. In order to set premiums, insurers review information such as age, gender, mileage, and automobile type. However, thanks to IoT, it is now conceivable to analyze information such as driving speed, driving duration, number of hard braking, drivers' traditional routes, and drivers' mobile phone use.

Automatic and smart manufacturing, which is proposed by Industry 4.0, requires cyber-physical systems (CPS) as its bedrock. CPS are systems that employ smart technology in all areas of production. These pieces may include machinery, cars, routes, merchandise, and the factory structure itself. The data is stored as well as secured by these highly sensitive sensors, which pass this information to the computers to analyze and make decisions. The decisions made by these computers are fed to the machines and implemented in the industry to better the manufacturing process with the help of these smart systems [20]. The integration of CPS also helps to improve the robotic machinery that is used in the new industry. These robots function in full automation without the requirement of any human interaction, and work only pertaining to the area they are deployed to. The CPS that are employed take the data and insights from all the areas of the manufacturing plant, thereby keeping all the machines coordinated and well within the loop. All the data that is collected from within the business is analyzed with the help of cloud technology to improve the output and coordination of the factories. Businesses may also employ AI and ML for better optimization based on prior outcomes [21]. CPS and IoT complement one other to develop smart manufacturing. Factories have decreased downtimes, improved efficiency, generated better goods, and achieved greater production. Factories that utilize IoT in production cite cost savings and an increase in quality.

Maintenance is a huge burden for plant management. The downtime and expenses connected with it might be costly. The standard method for routine servicing is unproductive and raises the danger of the breakdown and degradation of industrial machinery, which in turn raises expenses. Predictive maintenance solutions leverage IoT to obtain real-time information about all in-service equipment. Depending on the info, the system forecasts the time frame for maintenance operations. Manufacturing facilities feature linked systems where various forces are in action. Loading, design, and process changes at one place may influence the whole facility. IoT-based proactive maintenance and cloud technologies find considerable utility in such scenarios. The system collects data from assets throughout the facility to estimate maintenance needs. Smart maintenance and management systems using IoT may incorporate AI and ML as well. Together they examine the influence of all the processes on production to produce more accurate forecasts over time.

3.6 BLOCKCHAIN

Blockchain, when defined in layman's terms, can be stated as the technology or database that is immutable, and used to keep records as well as the assets of any business over a network [22]. An asset might be material (a home, vehicle, currency, property) or immaterial (proprietary information, trademarks, licences, branding). To be simpler, anything that holds monetary value can be placed in the blockchain to keep a record of it as well as exchange it over the blockchain network. This helps in reducing misplaced records as well as saves in costs associated with the asset. Blockchain [23] is a public and distributed system that employs encryption to link information, such as transactions, and keeps it in a common ledger. It is a system that comprises of an ever-growing collection of data called "blocks" that are joined by encryption to protect the security of data exchanged, payments, etc. Blockchains store data via a network of devices, making them not just decentralized but also distributed. [24] This suggests that no individual, organization, or anybody who owns the network may be able to modify it, but anyone may utilize it, contributing to its functioning [23].

The intricate relation of Industry 4.0 and blockchain is yet to be fully grasped and the meaning of the relation is yet to be fully understood. However, there are several situations in which blockchain delivers possible advantages. Many of us have heard of uses in monetary operations where confidence may be generated from the usage of blockchains [25]. Currencies and fiat money difficulties are avoided and there may be a regulated continuity in supply transactions when various layers in the supply base are engaged. These are all fantastic techniques of organizing transactions. There are a few additional sectors where blockchain may provide value in Industry 4.0. One of the locations might be where the blockchains incorporate product, assembly, or element-identifiable details. This might aid in measuring and addressing quality concerns with a better degree of precision.

Other areas where blockchain may provide value in Industry 4.0 might be connected to the goods themselves and their manufacturing or element identification information. This might be an indication of excellence with a greater degree of precision. For example, during recalls when the ability to precisely identify items showing the fault, blockchain might be particularly helpful. Here, blockchain may secure all the product information: its subassemblies, elements, allocation route to market, etc. Having access to the group of statistics at every point in the supply chain may decrease the cost and inconvenience linked to recalls. The option to monitor product details for companies that demand the utmost accountability, for example, medical or defence items, is particularly helpful. Blockchains may be formed using the new information acquired by video surveillance. This offers us access to relevant data than could reasonably be acquired by a person in a short length of time.

Another possible use is in leveraging the intrinsic encrypting of a blockchain. This might theoretically shield some of the data delivered over a public network from unwanted inspection or alteration. Although there are certain precautions that are handed out to prevent the misuse of information, there still leaves a small hole in the data security that is attributed to the user using a public network to transmit data, which can be easily breached and consequently disclosed in that manner.

In the intellectual property (IP) domain, blockchain may be utilized in various ways: proof of the licences, registration, and clear rights; retain records; control and monitor the distribution rights; build IP contracts; and to handle the monetary operations to acquire the rights. This might be particularly advantageous to SMEs to protect their ideas, since blockchain could provide better simplicity, transparency, and fewer middlemen to the patent ecosystem.

3.7 INDUSTRY 4.0 IN DIFFERENT SECTORS

The major advantage of implementing Industry 4.0 is that it provides real-time information regarding the whole manufacturing and assembly process. Even the economic cycle of the company is monitored by these technologies, e.g., the resources used in manufacturing, the availability of the resources via various phases, the derivation of the resources, and different processes related to the products. With the ability to keep record of the supply chain as well as all the production processes, a multinational corporation can easily make plans and protocols for controlling production processes and enhancing the productivity performance of the organization [26]. The allocation of materials and items may be handled efficiently to successfully regulate the production operations. These are the prime reasons that have heralded the technology of Industry 4.0 allowing for industries to up their level of efficiency and work smoothly. The rise in productivity may also allow a company to fulfil client requests in various market sectors.

On the flip side, in the manufacturing industry there is always scope for excess production, which may lead to wastage of resources, but these problems are readily solved by the usage of new age technologies. Industry 4.0 has also paved a simpler way for the business to consumer from of industry. Augmented reality is another integral part of Industry 4.0. Major firms utilize augmented reality gadgets like headsets and eyewear for constructing smart factories since they consider that the integration of virtual environment in intelligent production process may boost relationships between staff [27]. We also see that with the proper analysis of data, there can be a boost provided to the efficiency of the workforce. With the growing uses of these kinds of devices that stimulate the artificial world, workers are quick to adapt to them to further their growth in different industries.

Maintenance workers may readily look into machines by employing this augmented reality equipment. With the help of these kinds of new technologies, workers can easily gauge what is wrong with the machines. These new age gadgets allow people to go beyond the traditional ways of detecting problems allowing them to see beyond the man-made walls and other opaque objects, and with the use of it, it becomes easier for workers to decide where they must put in a hole or where to do the necessary repairs saving a lot of hassle. The use of robots is another highlight of this new era of technology. Robotics technology can conduct repeated activities in the industrial process and the risks of human mistakes also stay low. Human resources may concentrate on more creative and motivated areas of company. Given all these factors, it can be claimed that the deployment of Industry 4.0 is becoming important for major industrial enterprises.

The relevance of robotic technology in production is continually expanding because of the technological age of the 21st century for regulating their production rates. However, small organizations have not demonstrated adequate enthusiasm in embracing the new age technologies. But the declaration of COVID-19 as a pandemic in 2020 pushed these and other companies to digitalize most of their workforce and workload. During the lockdown period, many companies shifted to online work mode. However, some small companies were not able to adapt fully to the changing environment and that threatened their existence [28]. As a consequence, it has become extremely vital to deploy digital solutions in the industrial industry to control production levels so that the firms can fulfil the demands of target consumers in various market sectors.

Due to such magnanimous shifts in the daily lives of the people after the pandemic, the part played by the e-commerce industry rose to an unimaginable height. It became the nerve centre of industry, which allowed keeping track of all the production and deployment processes as well as managing all the resources and using them judiciously. The integration of these new technologies, which is driven by the network as well as the IoT-based gadgets, helps the industry to create a manufacturing process that is fully digital from the beginning to the end as well as to fully optimize the production process and rates. Many enterprises have adopted this new age technology to improve their manufacturing processes as well as to get a boost in their efficiency rate. Besides this, the application of Industry 4.0 also assists in monitoring and controlling the distribution of created items in various specialized markets.

Automation technologies and specifically the utilization of robots may assist in successfully executing repetitive activities in industries. In this way, the personnel cost of organizations drops, and the productivity of organizations rises.

3.8 HEALTHCARE INDUSTRY

Already, in several industries, technologies like AI, IoT, 5G, and other tools have a huge influence on the way we live and work. Biopharmaceutical businesses are increasingly learning that the new age technology helps provide better care to patients. The advent of the digital revolution and Industry 4.0 will open the way for exciting new prospects for organizations around the globe.

AI is the most prominent feature that has been discovered with the inception of this new age industry. For instance, AI can review millions of photographs and papers regarding a given ailment and automatically detect patients presenting indicators of that condition. AI is already finding its way into medical imaging equipment, allowing clinicians to rule out ailments and diseases more precisely. Additionally, adopting wearable devices paired with AI and ML technologies make it much simpler to perform large-scale clinical research over a long duration. AI can then swiftly assess the information and detect causes, consequences, and symptoms that humans might overlook. This might be a huge contribution toward curing illnesses that go unnoticed or are difficult to identify.

At the same time, using AI, clinicians are discovering more about how to manage their offices and practices more effectively. For example, AI tools may aid in more effective resource planning and management, discovering unused equipment, employees, or inefficient production scheduling. Smartwatches have already made an impact to the existing healthcare scene by letting consumers measure their activity or basal metabolism on a jog. However, there are more to wearables than consumers think. In the medical sector, these gadgets may assist physicians and nurses in following a patient's status wherever they are.

For example, Apple developed a movement disorder API a few years ago that facilitated physicians to collect more information on Parkinson's disease by tracking smartwatch users. Doctors may also put sensors in patient beds in a hospital to measure things like heart rhythm and pulse rate without continually checking in on each individual.

The development of smartwatches and sensors has heralded a new age of healthcare, where physicians and experts can offer treatment to patients who can't physically visit a hospital. When a patient wears a smartwatch, there may not be a need for them to travel to the hospital to, for example, get a proper consultation for surgery.

Figure 3.3 shows how wearable devices have boomed over the last few years and how their data is helping us understand each individual and well as keep proper track of their health-related activities and their history of chronic diseases. This huge amount of data is used by computers, coupled with big data, to analyze whether a person has serious issues, or if it is just a false alarm.

Also, patients may supply their physicians with an informed update by utilizing an application on a smartphone. This might also be potentially lifesaving in remote

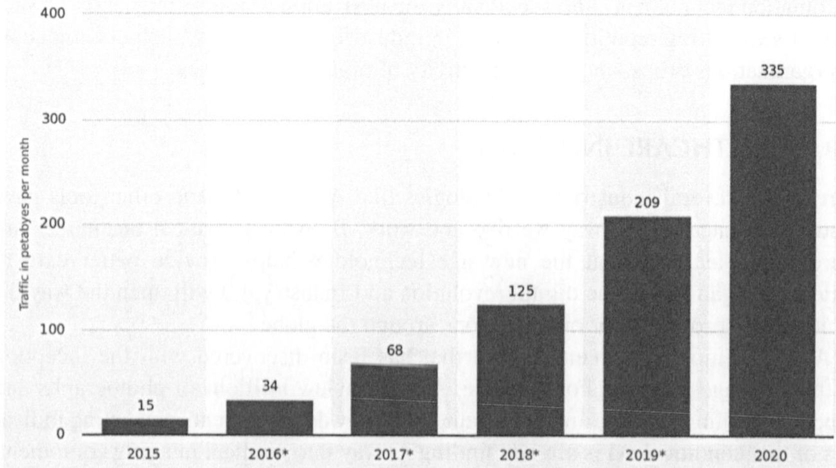

FIGURE 3.3 Global data of wearable devices. From Cisco Systems [29].

places where access to healthcare is restricted. Using data and insights obtained by smart wearables and then processed by AI and ML, healthcare practitioners may notice changes in an individual's health and swiftly react by arranging a preventative check-up or sending out medicine prescriptions.

3.9 INDUSTRY 4.0 CHALLENGES

When we adopt this new age industry in our daily lives, it introduces different technical obstacles, which creates ripple effects in every industry including the production industry. Thus, before integrating these technologies into our daily supply chain, it very important to create a well detailed and specially curated plan for every aspect of the industry, to achieve an agreement on safety and reliability problems, and to gather the required architectural expertise before taking such a huge step. Moreover, several experts argue that adopting Industry 4.0 is a challenging task for current industries, as there is a short turnaround for people to completely understand and get accustomed to the technology. New, innovative manufacturing methods include numerous facets, and confronts various sorts of obstacles, including and not limited to, those that are industry related, technical and financial hurdles, and societal and political problems.

At the beginning of the new age technology and industry, the driving factor for many new players was the ability to accept these new things as well as have proper knowledge regarding the people, e.g., innovative thinking abilities, critical challenges, and the capacity to keep up with the constant and dynamic changes as well as the new categories of jobs that emerged. Companies must also be ready to experiment with the mix and match of new technologies with the tasks they have at hand, which includes them siphoning off data and making the most of it during the various manufacturing processes [30]. The industrial society is changing rapidly and the demands of the masses have changed drastically, therefore we see Industry 4.0 as leading the way and catering to everyone. The implementation of Industry 4.0 to the current industry will lead to multiple changes in all the sectors, which helps people to create new business models and new methods of furthering the scope of industry.

Other obstacles and concerns of businesses are connected to innovation, technology components, digital transformation improvements, and the expanding interconnection developments, which play a significant function in every company. Some other obstacles that hamper the growth of Industry 4.0 pertains to the fact that there are a lot of technical competency issues, the digital changes that are rapidly increasing, and the networking issues. Companies must be able to fully adapt to these new shifts. As we have already discussed, the new industrial revolution comprises of delivering better and efficient manners of production that is strongly related with the full automation and robotization of all tangible resources and the incorporation into electronic environments of all distributors and suppliers [31]. The majority of businesses, particularly large and micro industries, are actually reluctant toward the digitalization process and the indecision is related to the fact that there are many barriers as well as very minimalistic progress in terms of the output. On the other hand, there is a lot of confusion with the acceptance of technology as it inculcates new threats as well as possibilities to the businesses. Traditional and small-business owners sometimes fear the loss of jobs and the lack of experience of trust of these new technologies. Also, the constantly changing dynamics in the industry and the novelty of the technology makes it very difficult for the people to put their faith in it. Although the society is equipped with the ability to make Industry 4.0 practical such as broadband technologies, accessible IoT gear, standardized communication protocol, gathering relevant data, and evaluating for implications are still major difficulties for Industry 4.0.

3.10 INDUSTRY 4.0 BENEFITS

We have seen throughout the chapter that the inclusion of such new technologies such as AI, big data, cloud computing [32], and IoT will change the course of industrial development, with more improved and sophisticated ways of running businesses as well as providing new models to establish an industry. As the industry becomes more digital, there are fewer issues in terms of mishandling and human error. When we look into the architectural aspect of the new industry, there are various advantages in terms of structural integrity and revenue generation. In reality, virtual reality and augmented reality prototyping give a thorough grasp of product features and advantages, permitting the dynamic study of the whole products functions among every person involved with the manufacturing process. Industry 4.0 has changed the way of business by providing new methods to conduct business as well as better sources to procure resources, changing the ways of the traditional manufacturing units. One of the reasons that has led to the boom in the usage of technology can be attributed to the fact that data is constantly growing, and making sense of it and using it judiciously can only be done with the help of machines. Businesses have started to look toward data as a valuable entity that can help them to peer inside their workings to a great extent. This importance of data makes companies understand that they must change the way they look towards data and their capability to handle the enormous amount of data that is being poured over onto them day by day. This becomes the biggest hurdle for the survival of conventional manufacturing firms [33]. Utilizing interactively programmable manufacturing tech in combination with enhanced adaptability of the device itself has numerous positive points, which are listed as follows, but are not limited to, custom products pertaining to each user, more robust distribution of assets among the industry, as well as alleviating of the demand–supply gap, which occurs for various reasons. This enables quicker, more economic, simpler, and more diversified manufacturing processes. Industry 4.0 provides various advantages for organizations in multiple dimensions. Industry 4.0 offers significant advantages, for instance, the decrease of wage costs, rationalization of business operations, minimization of inventory errors, and more transparent logistic activities. All of these are essential to enhanced production and income, which may, consequently, boost economic development.

3.11 CONCLUSION

After going through so many use cases of the new technology in this new industrial revolution, we can easily conclude that this is the need of the hour, and adapting these technologies will not only help us in terms of manual labour but also in terms of money and efficiency. These technologies can act as the first line of defence against unforeseen circumstances and then can act intelligently without any human intervention. This kind of behaviour would help industries to grow and flourish in accordance with the changing environment. Using these technologies we have become smarter in terms of healthcare and making well-informed decisions with the help of these technologies. There are still many hurdles that need to be crossed for us as a society to fully adapt to these new rules and technologies, but it is guaranteed that adopting new technologies would surely benefit us. We must strive to incorporate these technologies in our lives and make the most of the opportunities provided.

REFERENCES

1. Maurya, A., & Sharma, V. (2022). Facial Emotion Recognition Using Keras and CNN. In *2022 2nd International Conference on Advance Computing and Innovative Technologies in Engineering (ICACITE)*, pp. 2539–2543. https://doi.org/10.1109/ICACITE53722.2022.9823480.
2. Mohamed, Mamad. (2018). Challenges and Benefits of Industry 4.0: An Overview. *International Journal of Supply and Operations Management*. 5. 256–265. https://doi.org/10.22034/2018.3.7.
3. Kumar, Shashank, Narkhede, Balkrishna & Jain, Karuna. (2018). Industry 4.0: Literature Review and Future Research Directions. Accessed from: https://www.researchgate.net/publication/329643564_Industry_40_Literature_Review_and_Future_Research_Directions

4. Vuksanović, Dragan, Vešić, Jelena & Korčok, Davor. (2016). Industry 4.0: The Future Concepts and New Visions of Factory of the Future Development. 293–298. https://doi.org/10.15308/Sinteza-2016-293-298.
5. Ali, M. A., Balamurugan, B., Dhanaraj, R. K., & Sharma, V. (2022). IoT and Blockchain based Smart Agriculture Monitoring and Intelligence Security System. In *2022 3rd International Conference on Computation, Automation and Knowledge Management (ICCAKM)*, Dubai, United Arab Emirates, pp. 1–7. https://doi.org/10.1109/ICCAKM54721.2022.9990243.
6. Manikandan, N., Ruby, D., Murali, S., Sharma, V. (2022). Performance Analysis of DGA-Driven Botnets Using Artificial Neural Networks. In *2022 10th International Conference on Reliability, Infocom Technologies and Optimization (Trends and Future Directions) (ICRITO)*, pp. 1–6. https://doi.org/10.1109/ICRITO56286.2022.9965044.
7. Basheer, S., Singh, K. U., Sharma, V., Bhatia, S., Pande, N., & Kumar, A. (2023). A Robust NIfTI Image Authentication Framework to Ensure Reliable and Safe Diagnosis. *PeerJ Computer Science*. 9. e1323.
8. Umachandran, Krishnan, Jurčić, Igor, Corte, Valentina & Ferdinand-James, Debra. (2018). Industry 4.0.: The New Industrial Revolution. Accessed from: https://www.researchgate.net/publication/329539981_Industry_40_The_New_Industrial_Revolution
9. Vyshnevskyi, Oleksandr, Liashenko, Viacheslav & Amosha, Oleksandr. (2019). The Impact of Industry 4.0 and AI on Economic Growth. *Scientific Papers of Silesian University of Technology. Organization and Management Series*. 9. 391–400. https://doi.org/10.29119/1641-3466.2019.140.32.
10. Mezina, Tatiana. (2020). The Impact of "Industry 4.0" On Economy and Production. 670–677. https://doi.org/10.15405/epsbs.2020.12.04.77.
11. Topleva, Silviya. (2018). Industry 4.0: Transforming Economy Through Value Added. *Asian Journal of Economic Modelling*. 6. 37–46. https://doi.org/10.18488/journal.8.2018.61.37.46.
12. Ajmani, P., Sharma, V., Samuel, P., Somasundaram, K., & Vidhya, V. (2022). Patient Behaviour Analysis and Social Health Predictions Through IoMT. In *2022 10th International Conference on Reliability, Infocom Technologies and Optimization (Trends and Future Directions) (ICRITO)*, pp. 1–6. https://doi.org/10.1109/ICRITO56286.2022.9964846.
13. Popov, Vladimir, Kudryavtseva, Elena, Kumar, Nirmal, Shishkin, Andrei, Stepanov, Stepan & Goel, Saurav. (2022). Industry 4.0 and Digitalisation in Healthcare. *Materials*. https://doi.org/10.3390/ma15062140.
14. Ur Rehman, Mati, Andargoli, Amir & Pousti, Hamid. (2019). Healthcare 4.0: Trends, Challenges and Benefits. Accessed from: https://www.researchgate.net/publication/342981091_Healthcare_40_Trends_Challenges_and_Benefits
15. Juyal, Vandana, Saggar, Ravish, & Pandey, Nitin. (2020). On Exploiting Dynamic Trusted Routing Scheme in Delay Tolerant Networks. *Wireless Personal Communications*. 112. 1705–1718.
16. Durgalakshmi, K., Anbarasu, P., Karpagam, V., Venkatesh, A., Kannapiran, B., & Sharma, V. (2022). Utilization of Reduced Switch Components with Different Topologies in Multi-Level Inverter for Renewable Energy Applications-A Detailed Review. In *2022 5th International Conference on Contemporary Computing and Informatics (IC3I)*, Uttar Pradesh, India, pp. 913–920. https://doi.org/10.1109/IC3I56241.2022.10073430.
17. Juyal, Vandana, Singh, Ajay Vikram, & Saggar Ravish. (2015). Message Multicasting in Near-Real Time Routing for Delay/Disruption Tolerant Network. In *2015 IEEE International Conference on Computational Intelligence & Communication Technology*. IEEE.

18. Facchini, Francesco, Pinto, Luiz, Digiesi, Salvatore, Mossa, Giorgio & Venturini, Glória. (2021). Industry 4.0 Enabling Technologies Application in Healthcare Areas – A Systematic Review. https://doi.org/10.14488/ijcieom2020_full_0012_37273.
19. Juyal, Vandana, Pandey, Nitin, & Saggar, Ravish. (2016). Impact of Varying Buffer Space for Routing Protocols in Delay Tolerant Networks. In *2016 International Conference on Communication and Signal Processing (ICCSP)*. IEEE.
20. Pushpalatha, N., Jabeera, S., Hemalatha, N., Sharma, V., Balusamy, B., & Yuvaraj, R. (2022). A Succinct Summary of the Solar MPPT Utilizing a Diverse Optimizing Compiler. In *2022 5th International Conference on Contemporary Computing and Informatics (IC3I)*, Uttar Pradesh, India, pp. 1177–1181. https://doi.org/10.1109/IC3I56241.2022.10072844.
21. Ahmad, S., Mishra, S., & Sharma, V. (2023). Green Computing for Sustainable Future Technologies and Its Applications. In Grima, S., Sood, K., & Özen, E. (Eds.), *Contemporary Studies of Risks in Emerging Technology, Part A (Emerald Studies in Finance, Insurance, and Risk Management)*, Emerald Publishing Limited, Bingley, pp. 241–256. https://doi.org/10.1108/978-1-80455-562-020231016
22. Juyal, Vandana, Pandey, Nitin, and Saggar, Ravish. (2016). A Heuristic Light Weight Security Algorithm for Resource Constrained DTN Routing. In *2016 IEEE International Conference on Computational Intelligence and Computing Research (ICCIC)*. IEEE.
23. Ali, Md Akkas, Balamurugan, B., and Sharma, Vandana. (2022). IoT and Blockchain Based Intelligence Security System for Human Detection Using an Improved ACO and Heap Algorithm. In *2022 2nd International Conference on Advance Computing and Innovative Technologies in Engineering (ICACITE)*. IEEE.
24. Singh, Anamika, et al. (2022). Blockchain: Tool for Controlling Ransomware through Pre-Encryption and Post-Encryption Behavior. In *2022 Fifth International Conference on Computational Intelligence and Communication Technologies (CCICT)*. IEEE.
25. Sinay, Juraj & Kotianová, Zuzana. (2018). Automotive Industry in the Context of Industry 4.0 Strategy. *Transactions of the VŠB: Technical University of Ostrava, Safety Engineering Series*. 13. 61–65. https://doi.org/10.2478/tvsbses-2018-0014.
26. Öztürk, Fatih, Köten, Hasan & Karagoz, Yasin. (2018). The Effects of Industry 4.0 on Quality and Productivity in Automobile Industry. Accessed from: https://www.researchgate.net/publication/330764366_The_Effects_of_Industry_40_on_Quality_and_Productivity_in_Automobile_Industry
27. Gruber, Franz. (2013). Industry 4.0: A Best Practice Project of the Automotive Industry. https://doi.org/10.1007/978-3-642-41329-2_5.
28. Gökalp, Mert, Kayabay, Kerem, Akyol, Mehmet Ali, Eren, P. & Koçyiğit, Altan. (2016). Big Data for Industry 4.0: A Conceptual Framework. 431–434. https://doi.org/10.1109/CSCI.2016.0088.
29. Kaliraj, P. & Thirupathi, Devi. (2021). Big Data Applications in Industry 4.0. https://doi.org/10.1201/9781003175889.
30. Adam, Khalid. (2015). Big Data Analysis and Storage. Accessed from: https://www.researchgate.net/publication/313400371_Big_Data_Analysis_and_Storage
31. Data Accessed from: https://www.top500.org/news/market-for-artificial-intelligence-projected-to-hit-36-billion-by-2025/
32. Data Accessed from: https://www.statista.com/statistics/292837/global-wearable-device-mobile-data-traffic/
33. Sharma, V., Balusamy, B., Thomas, J. J., & Atlas, L. G. (Eds.). (2023). *Data Fabric Architectures: Web-Driven Applications*. Walter de Gruyter GmbH & Co KG.

34. Singh, A., Dhanaraj, R. K., Ali, M. A., Balusamy, B., & Sharma, V. (2022). Blockchain Technology in Biometric Database System. In *2022 3rd International Conference on Computation, Automation and Knowledge Management (ICCAKM)*, Dubai, United Arab Emirates, pp. 1–6. https://doi.org/10.1109/ICCAKM54721.2022.9990133.
35. Raj, A., Sharma, V., & Shanu, A. K. (2022). Comparative Analysis of Security and Privacy Technique for Federated Learning in IOT Based Devices. In *2022 3rd International Conference on Computation, Automation and Knowledge Management (ICCAKM)*, Dubai, United Arab Emirates, 2022, pp. 1–5. https://doi.org/10.1109/ICCAKM54721.2022.9990152.

4 Industry 4.0: Features Adopting Digital Technologies in the Oil and Gas Industry

Nasurudeen Ahamed N., Amreen Ayesha, and P. Karthikeyan

CONTENTS

4.1 Introduction ... 70
4.2 Overview of Industry 4.0 Technologies ... 70
4.3 Blockchain ... 71
4.3.1 Varieties of Blockchains ... 72
4.3.2 Opportunities and Applications of Blockchain Innovation in the O&G Industry ... 73
4.3.2.1 Robotizing Leakage Detection of Gas Pipeline ... 73
4.3.2.2 Following and Tracing O&G Shipments ... 74
4.3.2.3 Robotizing O&G Production ... 74
4.3.2.4 Overseeing Life Cycle O&G Assets ... 75
4.3.2.5 Improving Billing and Payments ... 75
4.3.2.6 Functional and Compliance Audit ... 75
4.3.2.7 Discernment of Well Abandonment and Restoration ... 76
4.3.2.8 Gathering and Disposing of Waste at Oil Refinery Stations ... 76
4.4 Artificial Intelligence ... 76
4.5 Internet of Things (IoT) ... 77
Security Concerns ... 78
Management Support ... 78
Firm Size ... 78
Trading Partner Pressure ... 78
Information Intensity ... 78
Customer Value ... 79
Environmental Value ... 79
4.6 Cloud Computing ... 80
4.7 Big Data in O&G Industry ... 81
4.7.1 Big Data in Upstream O&G Operations ... 82

DOI: 10.1201/9781003348313-4

4.7.1.1 Big Data in Exploration 82
4.7.1.2 Big Data in Drilling 83
4.7.1.3 Big Data in Reservoir Engineering 83
4.7.1.4 Big Data in Production Engineering 83
4.7.2 Big Data in Downstream O&G Operations 83
4.7.2.1 Big Data in Refining 83
4.7.2.2 Big Data in O&G Transportation 84
4.7.2.3 Use of Big Data by Health and Safety Executive (HSE) 84
4.8 Manufacturing Execution System (MES) for O&G Industry 84
4.8.1 Milestones of MES 85
4.9 Conclusion 86
Abbreviations 86
References 87

4.1 INTRODUCTION

The oil and gas (O&G) industry [1] is the primary wellspring of energy on the planet, with an exceptionally perplexing creation and conveyance process. It stands apart for the economic change of the world, supporting the interest in heat, power, portability, and other fundamental petrochemicals of society. The assembling and conveyance process includes best-in-class advances at various levels. These levels are upstream, halfway, and downstream. The upstream segment incorporates investigation and creation processes, such as geographical studies, and inland and seaward boring. The halfway portion includes transportation; stockpiling; and advertising raw petroleum, flammable gas, and delicate items.The downstream portion transforms O&G into a final product (example: petrol).

The O&G industry has recently been damaged because of the coronavirus pandemic and expanding high market instability. Moreover, the push for carbon charges (calculating the supplementary fees associated with emissions of carbon) and greener and cleaner energy is supposed to result in a drawn-out decrease in petroleum derivative interest and utilization. Subsequently, it is essential to distinguish the difficulties of common O&G areas to accomplish a more beneficial and proficient method for keeping the O&G industry more competitive.

4.2 OVERVIEW OF INDUSTRY 4.0 TECHNOLOGIES

The term Industry 4.0, also known as the Fourth Industrial Revolution, was first used in Germany in 2011 during the Hanover Fair. The nine leading technologies associated with Industry 4.0 [2] are Internet of Things (IoT), DB analytics, cloud computing, additive manufacturing, augmented reality, autonomous robots, cybersecurity, systems integration, and simulations, as illustrated in Figure 4.1. An overview of Industry 4.0 technologies as they currently stand is provided in the following sections. Industry 4.0 technology's approach, architecture, and protocols are not covered in detail in this book chapter, but references are included for more information.

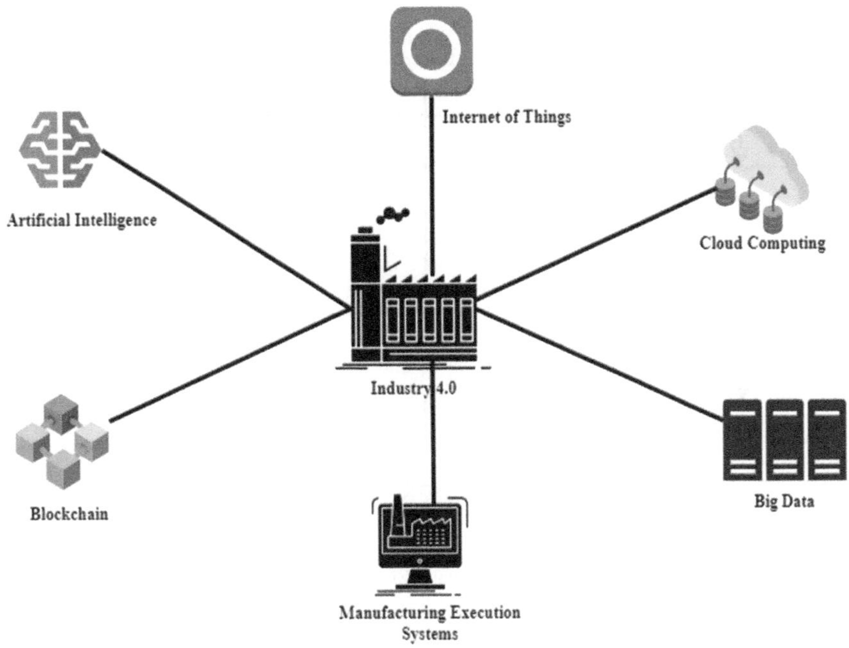

FIGURE 4.1 Technologies in the Industrial Revolution 4.0.

4.3 BLOCKCHAIN

The exchange of oil and gas items is a profoundly normalized and quality-delicate cycle that tends to be fully protected with quick information handling [3]. The segments of the O&G industry are:

- Upstream: investigation and creation of oil and gas.
- Halfway: coordinated factors, stockpiling, and transportation administrations.
- Downstream: managing the capacity and distribution of oil- and gas-based items like petroleum, diesel, and condensed flammable gas.

The more significant part of the frameworks created to screen and oversee exchange in the O&G industry is unified, inconsistent, and profoundly nontransparent. So, we can utilize one of the arising innovations: blockchain.

Blockchain is a record of exchanges that utilize cryptographic hash calculations, and makes the tasks of O&G clients profoundly productive, transparent, and trusted [4].

Blockchain can carry many advantages to numerous modern fields and it became an important device for Industry 4.0 applications. In private networks, standard data sets, as a rule, give a quick and fantastic asset for some applications (e.g. plumbing

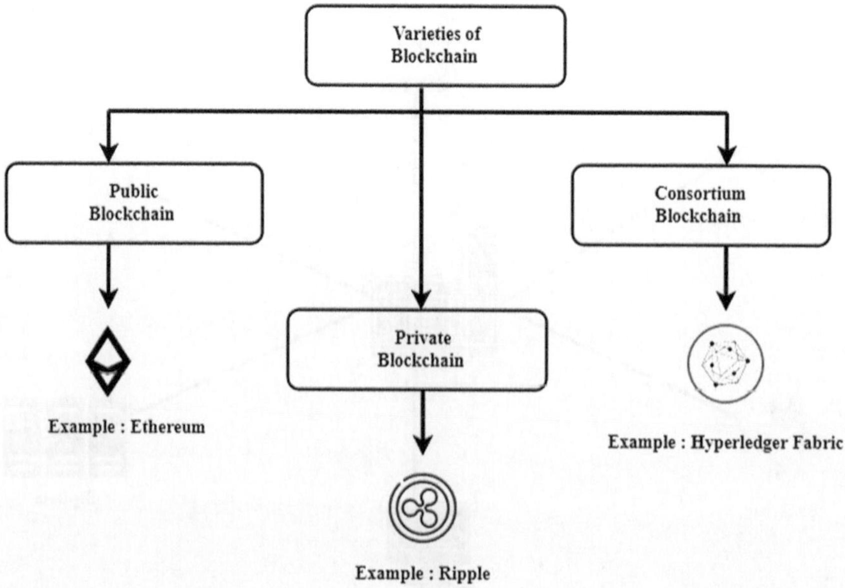

FIGURE 4.2 Varieties of blockchain.

and container cages). The various developments right now connect an equivalent requirement, as blockchain is not able to fulfil the needs of every example.

4.3.1 Varieties of Blockchains

(Also refer to Figure 4.2.)

- Public blockchains: Permission is not required to join the blockchain [5]. Anybody can distribute and approve exchanges. Examples: Bitcoin and Ethereum.
- Private blockchains: Support in the blockchain is controlled by a proprietor. The proprietor settles issues such as mining rewards, who can join the organization, or which hubs are assigned to go about as validators of the blockchain. Example: Ripple.
- Consortium or united blockchains: A hybrid of the first two types. A gathering of proprietors work on the blockchain. They restrict client admittance to the organization and the activities performed by members. Example: Hyperledger Fabric.

The following discusses the Hyperledger Fabric platform in the O&G industry. The blockchain-based model (Figure 4.3) has been proposed to computerize the supply network and strategies for management's tasks in the Ecuadorian business. It has utilized Hyperledger Fabric [6] and Composer to foster the chain codes for carrying

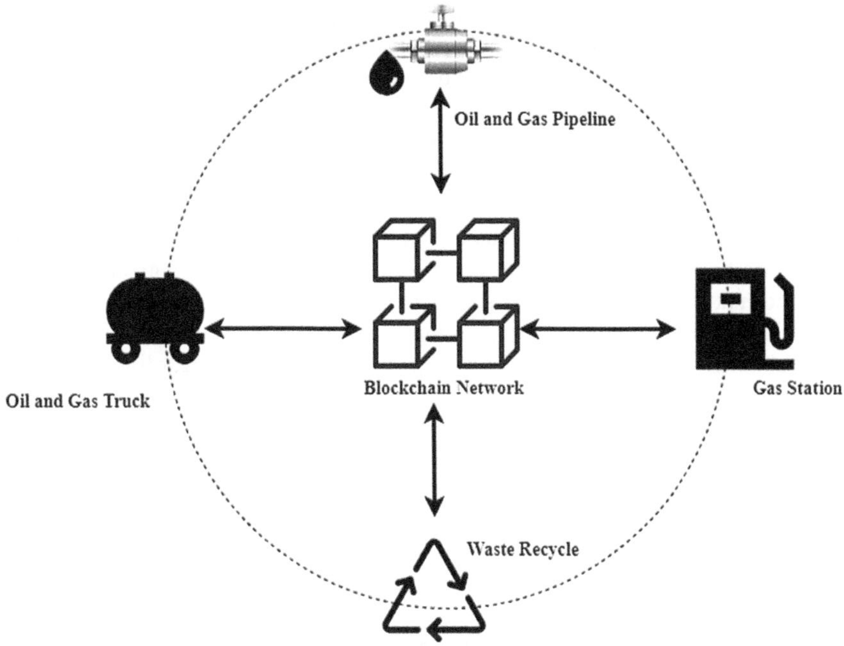

FIGURE 4.3 An Overview of the O&G industry integrated into blockchain technology.

out production networks and executive administrations for the O&G industry. The chain codes have characterized the guidelines observed during O&G operations to develop proficiency, efficiency, and directness. The vital partners considered by the proposed model incorporate gas providers, makers, merchants, retailers (fuel stations), and clients.

4.3.2 Opportunities and Applications of Blockchain Innovation in the O&G Industry

- Robotizing leakage detection of gas pipeline [7]
- Following and tracing O&G shipments [8]
- Robotizing O&G production [9]
- Overseeing life cycle of O&G assets [10]
- Improving billing and payments [11]
- Functional and compliance audit [12]
- Discernment of well abandonment and restoration [13]
- Gathering and disposing of waste at oil refinery stations [14]

4.3.2.1 Robotizing Leakage Detection of Gas Pipeline

The leakage of gas is an exceptionally important concern in the oil industry. Sensors are frequently joined to the gas pipelines to gather information about the state of the pipeline. Such data can aid in finding or anticipating a gas spillage, and the current condition and life expectancy of a pipeline. Even though sensors are useful, we cannot always prevent gas spillage, and a significant disaster could happen. Thus, this can influence the dependability, accessibility, and security of the information utilized for pipeline conditions and spillage recognition.

Blockchain is prepared to precisely recognize gas pipeline spillage and protect the condition of pipelines. It can help check that the pipeline-related activities of partner organizations, and maintenance and repairs are within safety regulations. Blockchain-based agreements can be created to help O&G organizations handle such information and find the gas spillage area in a pipeline. Through blockchain-based agreements, warnings can be promptly created and sent to pipeline administrators about any spillage issue, such as sharing of information regarding the location and needed repairs.

Unrefined petroleum is moved from the wellhead to the processing plant via pipelines, railways, trucks, and ships. The oil-based goods created at processing plants are frequently delivered to businesses or local clients through railroad tank vehicles, trucks, or pipelines. But, due to the high strain and explosion dangers, giant haulers are not preferred to deliver compacted flammable gas. It is first changed into liquefied petroleum gas (LNG), and then it is transported to clients.

4.3.2.2 Following and Tracing O&G Shipments

Following and tracing of O&G shipments allow members to distinguish, check, screen, record, and successfully deal with the information about the client's activities and results. Businesses can help clients check the authenticity of the information and guarantee the consistency of activities within standards and supply network tasks in the O&G area. Additionally, the ongoing area of O&G items and the condition of the cargo and transfer process can be checked, confirmed, and reexamined.

The decentralization aspect of blockchain innovation makes it unmodifiable by outsiders. All transporting updates and changes are quickly recorded on the blockchain. Such records can be utilized to lay out the origin of the goods manufactured from oil and natural gas to distinguish any fraud connected with O&G items.

4.3.2.3 Robotizing O&G Production

Blockchain innovation [15] can help businesses in robotizing payments to shipping organizations in a quick, straightforward, and correct way based on the equipment use rate. In customary frameworks, such information can be modified by outside or competing businesses or programmers. But, blockchain guarantees that information created by sensors is unchangeable. For example, the amount of information from seismic studies is huge; such information can be safeguarded by producing and organizing the information on the blockchain stage (from manufacturer to consumer). Likewise, through confidential blockchains, the proprietary innovations of O&G businesses alongside other sensitive information, for example, well logs, can be safeguarded.

4.3.2.4 Overseeing Life Cycle O&G Assets

Blockchain can help determine the O&G equipment that requires upkeep, repair, and operations supervision based on their presentation score. Blockchain can record the upkeep history of O&G equipment, such as breakdown hardware ID, repair station ID, repair date, well-being score of broken down hardware, and life expectancy. The entire upkeep history of the equipment can be helpful to guarantee consistency with safety and maintenance regulations for reservoirs' development changes, reclamation, deserting, and oil creation tasks [16].

4.3.2.5 Improving Billing and Payments

In the production network, the O&G activities and many kinds of billings and payments are made and divided between various businesses. The O&G inventory network board cycles, for example, obtainment of raw petroleum by an organization and refinement of unrefined petroleum, require making, putting away, refreshing, and dealing with numerous billing records. It can take as long as 90 days to finish the accounting cycle in the O&G industry. The primary drivers of such delay are reliance on a unified framework that is helpless against hacking, human intervention for business contracts, and no transparent frameworks [17].

Blockchain backing of digital money engages an O&G business to make micropayments to the accounts of other partner companies; subsequently, it offers a secure and straightforward framework that does not include middlemen to settle payment-related disputes. Blockchain-based agreements can be created and sent to the participating associations to store an escrow sum (two times the product cost) in the smart contract wallet. Additionally, in different joint endeavours of O&G organizations, the trusted blockchain can be utilized to confirm, ascertain, and remit invoices.

4.3.2.6 Functional and Compliance Audit

The current frameworks used to deal with the information fundamental for compliance needs to give information about permeability, security, protection, and adaptation to internal failures [18]. Blockchain innovation can help state offices and experts follow the distributed guidelines to protect the people and climate. The security, supply network detectability, and unmodifiable information advantages of blockchain can assist specialists to recognize the sort of synthetic compounds utilized in the water-powered fracking process. There are many synthetic substances available to perform pressure-driven fracking. However, a few are dangerous to human well-being, and their utilization is internationally restricted. With blockchain records, compliance experts can get to the information and exchanges to confirm that an O&G company does not utilize prohibited synthetic substances. Authorities can issue penalties to O&G businesses that use dangerous and unacceptable synthetics during fracking.

4.3.2.7 Discernment of Well Abandonment and Restoration

Abandonment and restoration of wells is the last phase of the O&G improvement process [19]. The temporary well-abandonment type includes removing well control equipment from a drilling rig, and guaranteeing that the O&G extraction cycle will not continue. Abandoned O&G wells sites that are never being reutilized or returned require reclamation. When an O&G well is abandoned, it is guaranteed that the repositories are appropriately plugged to stop the stream into different supplies. After the abandonment of an O&G well, the site is reestablished to the previous state to limit the natural impacts of the remediation process.

Blockchain technology enables the storage and verification of various information related to oil wells. This includes details like the well owners, reasons for well abandonment, type of abandonment, confirmed or estimated oil reserves, and security measures taken for site remediation. This information is valuable for authorities as it helps them validate records and activities. Furthermore, it can be utilized in proposals for repurposing the surrounding land, such as converting it into farmland [20]–[21].

4.3.2.8 Gathering and Disposing of Waste at Oil Refinery Stations

Blockchain innovation helps clients limit unlawful unloading and disposal of sludge by storing the exchanges and information on the circulated record. Smart contracts can aid in fundamentally limiting waste-related delays, prevent records from being modified, and guarantee high information consistency [22]. The following elements can help businesses confirm the movement trucks delivering waste. The immutable exchanges, information sources, and smart contracts work on the trust between the partners engaged in waste disposal in the O&G industry. Blockchain additionally provides convenient, secure, and occasion-based moving of plans of the sludge authorities, self-regulators, or transporters.

4.4 ARTIFICIAL INTELLIGENCE

Artificial brainpower (computer-based intelligence) [23]–[24], a universally applicable innovation, is quickly entering many sectors. In particular, the O&G industry's artificial intelligence (AI) made significant advancements and changed the rules of the game.

As O&G organizations are much faster in embracing innovations than exploring different avenues regarding and changing their plans of action, the essential objective of AI (and other digitalization) endeavours are to further develop effectiveness. Practically speaking, this means regularly speeding up cycles and lessening risks.

For example, drilling (upstream) to utilize AI calculations. Its methodology depends on the natural human framework.

Here we discuss Artificial Neural Network(ANNs) in the O&G industry. In previous years ANNs proliferated. ANNs are utilized to foresee boiling point of unrefined petroleum, decide water–oil relative permeability of supplies, and produce hydrocarbons from the shale repository.

The ANN strategy estimates porosity tension in the iron arrangement. A boring machine was recently set up with AI, image processing, and ANN. In the initial step, a general model was found and adjusted to picture handling programming so that any shape that comes naturally could be tried. By setting up the machine in this way, any results can be sent to the boring control unit.

In order to reduce the expense of manufacturing, drilling equipment that include AI and image processing using ANNs are beneficial, leading to a programmed machine station. The erroneous tedious interaction should give the device a distorted kind of outcome. It utilized three divergent sorts of boring apparatus in preliminary penetration testing. This was because the fundamental point was to foster an insight machine. Thus, device twisting was not as expected, yet delivering AI boring machines for an enormous scope has been considered.

4.5 INTERNET OF THINGS (IOT)

IoT enables machine-to-machine (M2M) communication over an organization without requiring human–PC communication (Figure 4.4) [24]–[25]. The machine comprises an installed framework with sensors/actuators that send information over the web utilizing different communication channels. M2M is made conceivable by maintaining resource records, which permits gadgets to connect through characterized communication conventions and models. IoT has developed throughout the years with special emphasis on different areas. This study proposes recommendations for each factor based on importance–performance map analysis (IMPA) results. Research recommendations [26]–[27] were assessed by three senior managers from various multinational O&G companies. The evaluation accorded the suggested techniques for the following factors a high relevance rating:

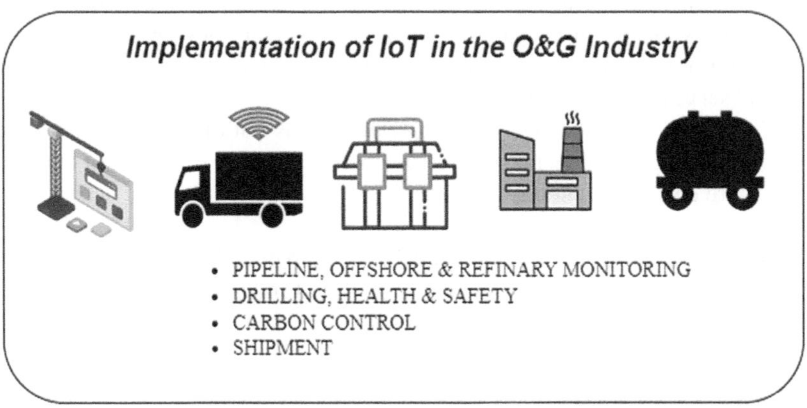

FIGURE 4.4 IoT in O&G industry.

SECURITY CONCERNS

- Strengthen cybersecurity defences against potential assaults on the online environment for oil and gas.
- Implement improved key management, negotiation, penetration testing, and end-to-end authentication techniques in network layer security rules.
- Implement data security and privacy policies at the application layer.

MANAGEMENT SUPPORT

- Ensure that upper management receives accurate and complete information for decision-making.
- Establish active and efficient involvement of senior management during implementation.
- Make the company's goal to adopt IoT with a high-level strategy roadmap supporting the company strategy and the core business [28].

FIRM SIZE

- Establish clear and comprehensive information that will be shared to raise worker awareness. Establish the workforce's active and efficient participation during implementation.
- Ensure the company is taking advantage of and using multiple levels of internal resources.

TRADING PARTNER PRESSURE

- Establish a network of cooperation with the trade or commercial partners, sharing potential expenses and risks associated with adoption of operations, gaining access to new technology, utilizing complementary skills, and exchanging information capacities to affect adoption outcomes.
- Hold a workshop on exchanging technology with business or economic ties, including a strategic partnership, to indirectly impact business development and serve as a source of IoT.
- Developed cross-managerial level relationships with trade or commercial partners regarding technology deployment to develop commercial trade to ensure enduring relations.

IoT has prepared for a few developments in the market and commenced the idea of Industrial IoT (IIoT). IIoT assumes an essential part in the industry by giving proficient and effective analysis frameworks that lessen expenses and increase efficiency.

INFORMATION INTENSITY

- Examine Industry 4.0 concepts for process automation to control the amount of information in the business after technology adoption [29].

- Examine ways to process a large amount of data already existing in the company using advanced analytics, big data, and information science.
- Restructure the organization's data infrastructure network, connectivity, and accessibility so that all are parts of an information infrastructure support information intensity.

CUSTOMER VALUE

- Promote a technology-adopting environment with a significant customer focus and be receptive to new concepts and methods of operation.
- Evaluate the value impact on customers, including potential benefits from a business-to-consumer (B2C) standpoint, such as savings in terms of time and costs, as well as before IoT deployment.
- Create ongoing initiatives and value additions or prospective financial impacts of IoT innovation to foster and uphold connections and confidence.

ENVIRONMENTAL VALUE

- Before deploying IoT, do an evaluation and analysis phase of the environmental effects of value, including pollution, water consumption, and spillages.
- Establish company-wide guidelines to support the transition to a low-carbon economy and a reform plan for more sustainable, resilient, and equitable growth utilizing the new IoT technology [30].
- Invest in human resources department and training initiatives that foster innovative digital thinking and produce a digitally purposeful company that prioritizes capturing the environmental benefits.

Figure 4.5 shows a usable IoT engineering for the upstream O&G segment. It comprises of layers of communication innovation, implementation phase, and core network. The actual layers comprise IoT hubs that gather information and control O&G gadgets and offices. The hubs send information to gateways or base stations utilizing different communication technologies. The communication layer incorporates small advances that can be sorted into short-range and long-range exchanges. Instances of short-range communication innovations incorporate Wi-Fi, Bluetooth, and Zigbee, while long-range communication incorporates low-power wide-area (LPWA) advances, for example, narrowband IoT, LoRa, and Sigfox. Unlicensed far-off innovations (LoRa and Sigfox) benefit far-off regions without cell inclusion. The network layer incorporates innovations such as distributed computing, programming characterized networks, blockchain, and network servers [31]–[32]. Information from the hubs is directed safely through different organization layer advancements to the application layer. The application layer enables information handling, information examination, and perception. IoT advances, conventions, and other related phrasing are widely discussed. IoT can be applied in tasks like control and observing of activities, foresight, computerization and control, well-being, and security of the O&G industry.

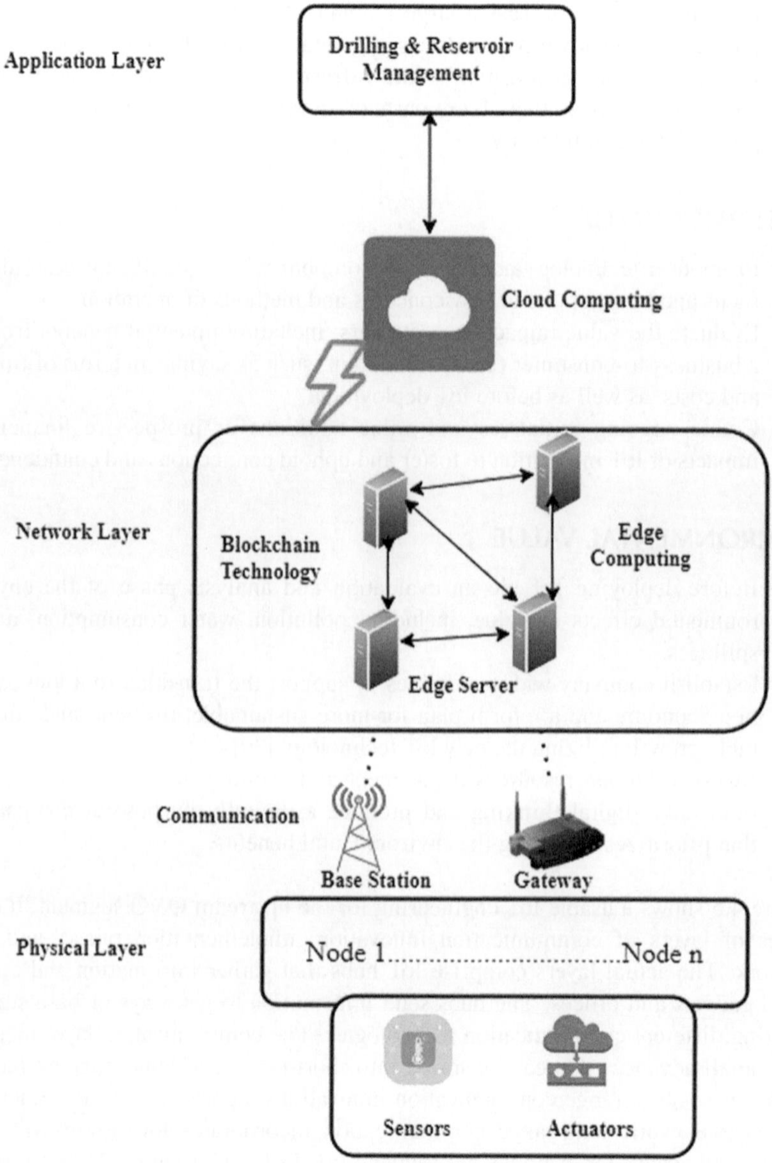

FIGURE 4.5 IoT Architecture for O&G industry.

4.6 CLOUD COMPUTING

Cloud computing is an integral part of Industry 4.0 because of its many benefits to businesses and institutions. It includes using cloud-based services provided on demand, such as data centres, networks, software, and intellectual capacity. Cloud technologies [33, 34] may reduce costs and increase income, improve safety and

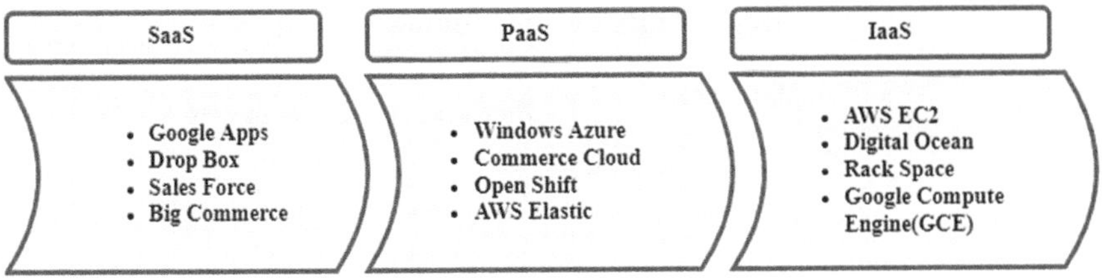

FIGURE 4.6 Cloud Services in the O&G sector.

performance, and increase efficiency and agility. IaaS (infrastructure as a service) provides pay-as-you-go services for networking, virtualization, and storage. PaaS (platform as a service) offers a software development platform delivered over the internet. SaaS (software as a service), IaaS, and PaaS can benefit from mobile access to online software, scalability, and reduced hardware costs. Different cloud computing services are shown in Figure 4.6. Although cloud computing offers many advantages, there are some known limitations, including reduced service quality due to snags in applications that require quick responses. Therefore, this is being researched to integrate cloud services with other innovations such as edge/fog computing.

4.7 BIG DATA IN O&G INDUSTRY

A vast amount of structured and unstructured data is generated in the O&G industry. Figure 4.7 depicts the characteristics of big data. The algorithm [35] is applied to the O&G industry big data and generates useful information (Figure 4.8).

The volume of data during drilling operations is another remarkable development in drilling tools. Tools such as measuring while drilling (MWD) and logging while drilling (LWD) are used to deliver real-time data. To support parameters such as fluid pressure, temperature, and composition in the manufacturing of O&G, optical fibres, in combination with various sensors, are now widely used. The concept of

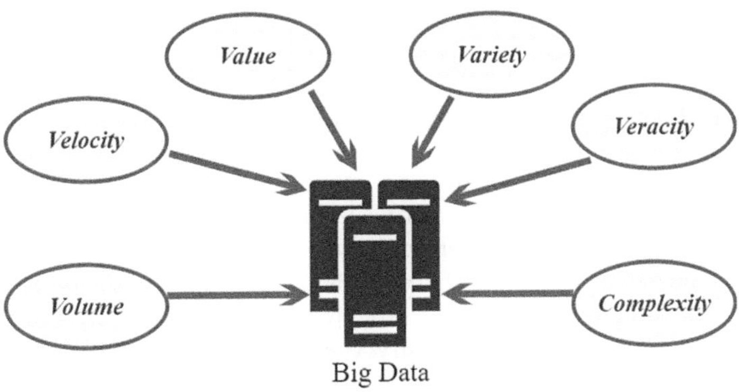

FIGURE 4.7 Characteristics of Big Data.

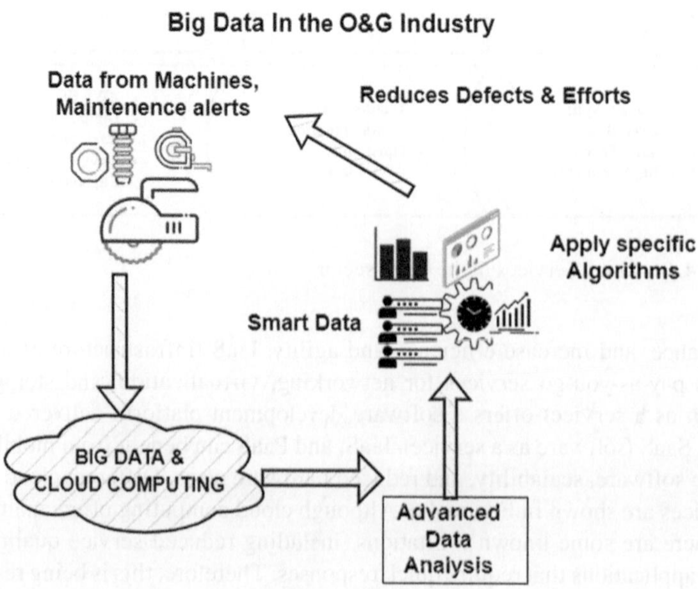

FIGURE 4.8 Big Data in the O&G industry.

speed as a characteristic of big data refers to the speed of sending and processing data. It also refers to the fast pace of creating data. A challenging problem regarding the velocity component is the limited number of available processing units compared to the amount of data. The speed of data creation has recently been tremendous, for example, data on 5 exabytes were generated in just two days. This is generally equivalent to the amount of human data that was created by 2003. Due to the complexity of various petroleum technical challenges, the velocity feature is much more evident in the O&G industry. Processing such information collected by one individual to resolve a complicated problem is impossible, takes a long time, and is uncertain. The industry often requires rapid, real-time data processing. For example, rapid processing of good data during drilling can lead to kick identification and effectively prevent harmful spillage.

4.7.1 Big Data in Upstream O&G Operations

4.7.1.1 Big Data in Exploration

Seismic data analysis requires sophisticated computer processing and powerful visualizations. The volume of data generated has significantly increased due to new developments in seismic equipment. These new data sets should be analyzed in detail, extending beyond what is currently available. Seismic data analysis is a basic big data usage in the O&G industry. Especially in the current times of dealing with large data sets, machine learning algorithms can more effectively identify the relationship between recorded data.

4.7.1.2 Big Data in Drilling

There are several data sources in the drilling sector, but most are data produced by digital drilling rigs and manually entered by human operators. Various studies can be carried out from the plan through the actual drilling operation using the data obtained from multiple drilling activities. Big data techniques may now be widely applied in drilling operations thanks to the development and adoption of new data logging tools and formats. More than 60 different sensors now cover a range of drilling parameters. Using collected real-time records, big data is utilized to recognize invisible non-productive time (INPT). The authors increased drilling operations by using cloud computing, artificial intelligence, and basic statistics to improve INPT. Drilling operation risks are reduced while using big data. Drill-string dynamics are computed using data from downhole vibration sensors.

4.7.1.3 Big Data in Reservoir Engineering

The authors improved drilling operations with cloud computing, artificial intelligence, and quantitative statistics to improve INPT. Drilling operation risks are reduced using big data. Drill-string dynamics are calculated from data by the downhole accelerometer sensor. This new technology has large amounts of data and artificial intelligence in integrated asset modelling (IAM) and closed-loop reservoir management. By forecasting influence parameters not available in theory-based equations of state, data-driven methods can improve model performance. The result is a new, data-driven approach for reservoir modelling.

In their study, Xiao and Sun made great use of hydrodynamic reservoir simulation to maximize the use of enhanced oil recovery (EOR) projects. Furthermore, the reservoir for a unique O&G well was modelled utilizing big data.

4.7.1.4 Big Data in Production Engineering

Electric submersible pumps (ESPs) have already seen improved performance due to big data adoption. Based on a three-phase process, big data is used to improve the performance of rod pump balloons. Shale operators use big data to improve hydraulic fracturing operations. The second step is automated workflows that carry out the necessary tasks. Interactive visual analytics is the third step and provides a user-friendly interface to capture the results. Big data is also employed to generate production-type AI-based curves to aid in the development of the industry.

This is the initial phase of a multistep data processing process that involves collecting raw data (both from structured and unstructured databases), filtering it, consolidating it with other data, and sending it to the machine learning pipeline.

4.7.2 Big Data in Downstream O&G Operations

4.7.2.1 Big Data in Refining

Historical data was once analyzed and processed in three steps to enhance petrochemical wealth management. A piece of equipment was a four-step cracked gas compressor (CGC). Using recent and historical operational data, the research first predicted CGC performance. The prediction of the CGC performance was established in the subsequent phase depending on the needs for the device's final moments and its failure circumstances. Finally, a user-friendly and visual report detailing the estimated CGC performance was provided for use in management decisions. These prognostic reports can significantly save stop and maintenance expenses because they were produced utilizing data analysis.

Management optimization is conducted using big data analysis for integrated refineries owned by Repsol in Spain. Google Cloud would offer Repsol data analysis products, advice, and services for this project, as well as automatic learning services. The workflow for researching the effects of completion parameters on good production was created using bulk product providers. They collected information from 4500 wells for slickwater processing. They investigated the impact of two distinct substances: flow-based flow aids and linear guar gels. Additionally, they accessed the IHS Energy database for statistics on monthly production. T-test was the statistical method utilized to assess the data.

4.7.2.2 Big Data in O&G Transportation

A study was done to employ big data analytics to enhance the performance of ships, predict propulsion power to enhance ship performance, and subsequently reduce greenhouse gas emissions. Sensors placed throughout large car and truck carriers (LCTCs). Measurement and verification (M&V) were used to collect the data over three months. Later, the data analysis was carried out using multi-layer perceptron (MLP) neural networks and extreme gradient boosting (XGBoost).

4.7.2.3 Use of Big Data by Health and Safety Executive (HSE)

An energy conservation model was developed using big data, and operational information gathered from marine operations. Scientists used data from publicly available automatic identification systems and information from the marine environment to generate an energy indicator. The fuel consumption of the ship's engines divided by the operation's weight and distance was used to define energy efficiency. It utilized the Hadoop framework and Apache Spark to implement big data.

The analysis of bulk product providers was used to improve safety at the work of O&G by managing the risk and improving safety. Data gathering and processing came first in the process. To do this, standard error measurement is used to eliminate the aberrant values from data sets after raw data had already been filtered according to the recording quality. The data is then structured via format conversion and decoding; the next phase involves regrouping and mapping the analysis to detect underlying, undiscovered tendencies. In the end, a multivariate statistical study was used to produce a result that was simple to understand.

4.8 MANUFACTURING EXECUTION SYSTEM (MES) FOR O&G INDUSTRY

Changing information into meaningful business information is crucial to improving and augmenting business potential. Today, a treatment facility business should be more reactive and receptive to fluctuations within the market. Effective functions of executives includes incorporating arranging, booking, execution, and the capacity to answer change promptly. This section investigates the foundation of manufacturing execution systems (MES) and uncovers how MES programming innovation conveys more proficient information to management, further develops creation execution, and improves functional execution, empowering superiors to rapidly transform information into benefits.

4.8.1 Milestones of MES

With excellent comprehension of how their plants act continuously, superiors can alleviate concerns with sensible, informed choices about creation execution. Conveying items that reliably meet client assumptions, assists purifiers with staying on task and accomplishing higher benefits. MES gives knowledge to progressively advanced tasks with fast, precise, and straightforward information.

There are three important achievements for MES, which structure the mainstays of productivity:

- More effective information for the board (MES 1.0: coordinating the past)
- Further developed creation execution (MES 2.0: the time of work process computerization)
- Improved execution of the board (MES 3.0: innovation progressing)

The oil business started to perceive the significance of robotized work processes. Essentially depending on information. During this period, programming was not available that permitted purifiers to computerize oil developments requests of handling and observing. Often, the planning capability sends down orders that specific oil developments need to happen. The traditional work cycle of oil developments had numerous manual advances, and successful checking of oil development tasks was put to the test.

Programming created during this period robotized oil development observation. It determined the gross volume in tanks utilizing, because of continuous assessments of tank level and characteristics. Before the development started, it guaranteed that there was sufficient source material and adequate space at the objective to finish the development satisfactorily. It likewise cautioned the administrator when a development ought to happen when 90% of the development had been reached, on the off chance that a tank was in twofold development and whether unapproved developments were occurring.

In this period, all protocols were recorded, consistently giving an accurate and ideal record of fluid streams and inventories and enabling more informed business choices. The computerized framework could accommodate creation on a day-to-day basis instead of the month-to-month premise, giving purifiers more certainty of actual stock positions, including the option to organize the specific stock in each tank.

4.9 CONCLUSION

This book chapter outlines Industry 4.0 using IoT, big data examination, distributed computing, additive manufacturing, augmented reality, independent robots, digital security, framework coordination, and reproductions, and the jobs they play in the upstream business O&G. A systematic literature review of Industry 4.0 innovations is introduced for a few tasks in O&G upstream operations. This incorporates investigation and improvement, diffusion and provision of final productspenetrating and finish, creation and enhancement, supply designing, control activities, and the activity of hardware and parts. A calculated system of Industry 4.0 for O&G upstream operations is introduced. We covered future patterns and opportunities and examination processes expected to join Industry 4.0 in the upstream downstream segments of the O&G industry.

ABBREVIATIONS

AI	artificial intelligence
AM	additive manufacturing
AR	augmented reality
B2C	business-to-consumer
BD	big data
CGC	cracked gas compressor
EEOI	energy efficiency operational indicator
EOR	enhanced oil recovery
ESP	electric submersible pump
ANN	Artificial Neural Network
IaaS	infrastructure as a service
IAM	integrated asset modelling
IoT	Internet of Things
IIoT	Industrial Internet of Things
INPUT	invisible non-productive time
In4.0	Industry 4.0
IPMA	importance–performance map analysis
LNG	liquefied natural gas
LCTC	large car and truck carrier
M2M	machine-to-machine
MLP	multilayer perceptron
O&G	oil and gas
PaaS	platform as a service
SaaS	software as a service
SCADA	supervisory control and data acquisition
XGBoost	extreme gradient boosting

REFERENCES

1. Spelman, M., Ashraf, M., & Weinelt, B. (2017). Digital Transformation Initiative–Oil and Gas Industry. In *World Economic Forum: Geneva.*
2. Gou, Q., Yan, L., Liu, Y., & Li, Y. (2013, August). Construction and Strategies in IoT Security System. In *2013 IEEE International Conference on Green Computing and Communications and IEEE Internet of Things and IEEE Cyber, Physical and Social Computing* (pp. 1129–1132). IEEE.
3. Wee, S. H. (2012, May). Important Enabler in the Knowledge Sharing Process: Top Management Support. In *2012 International Conference on Innovation Management and Technology Research* (pp. 657–662). IEEE.
4. Kalkan, A., Erdil, O., & Çetinkaya, Ö. (2011). The Relationships between Firm Size, Prospector Strategy, Architecture of Information Technology and Firm Performance. *Procedia-Social and Behavioral Sciences, 24*, 854–869.
5. Mori, J., Kajikawa, Y., & Kashima, H. (2011, June). Finding Business Partners and Building Reciprocal Relationships-A Machine Learning Approach. In *First International Technology Management Conference* (pp. 1069–1073). IEEE.
6. Melgin, T. (2015). Increasing Information Intensity in Industrial Services: Towards Industrial Internet and Industrie 4.0 Servitizing Industrial OEMs.
7. Balusamy, B., Sridhar, J., Dhamodaran, D., & Krishna, P. V. (2015). Bio-inspired algorithms for cloud computing: a review. International Journal of Innovative Computing and Applications 6(3–4), 181–202.
8. Satar, S. B. A., & Yusof, A. F. (2019, December). Exploring Internet of Things Adoption in Malaysian Oil and Gas Industry. In *2019 6th International Conference on Research and Innovation in Information Systems (ICRIIS)* (pp. 1–5). IEEE.
9. Rizwan, P., Babu, M. R., Balamurugan, B., & Suresh, K. (2018, March). Real-Time Big Data Computing for Internet of Things and Cyber Physical System Aided Medical Devices for Better Healthcare. In *2018 Majan International Conference (MIC)* (pp. 1–8). IEEE.
10. Elijah, O., Ling, P. A., Rahim, S. K. A., Geok, T. K., Arsad, A., Kadir, E. A., ... Abdulfatah, M. Y. (2021). A Survey on Industry 4.0 for the Oil and Gas Industry: Upstream Sector. *IEEE Access, 9*, 144438–144468.
11. Mohammadpoor, M., & Torabi, F. (2020). Big Data Analytics in Oil and Gas Industry: An Emerging Trend. *Petroleum, 6*(4), 321–328.
12. Wanasinghe, T. R., Gosine, R. G., James, L. A., Mann, G. K., De Silva, O., & Warrian, P. J. (2020). The Internet of Things in the Oil and Gas industry: A Systematic Review. *IEEE Internet of Things Journal, 7*(9), 8654–8673.
13. Balusamy, B., Krishna, P. V., Arasi, G. T., & Chang, V. (2017). A Secured Access Control Technique for Cloud Computing Environment Using Attribute Based Hierarchical Structure and Token Granting System. *International Journal of Network Security, 19*(4), 559–572.
14. Hawash, B., Mokhtar, U. A., Yusof, Z. M., Mukred, M., & Gaid, A. S. (2021, March). Factors Affecting Internet of Things (IoT) Adoption in the Yemeni oil and gas sector. In *2021 International Conference of Technology, Science and Administration (ICTSA)* (pp. 1–7). IEEE.
15. Ahmad, R. W., Salah, K., Jayaraman, R., Yaqoob, I., & Omar, M. (2022). Blockchain in oil and gas industry: Applications, Challenges, and Future Trends. *Technology in Society, 68*, 101941.
16. Gezdur, A., & Bhattacharjya, J. (2017, September). Digitization in the Oil and Gas industry: Challenges and Opportunities for Supply Chain Partners. In *Working Conference on Virtual Enterprises* (pp. 97–103). Springer, Cham.

17. Solanki, P., Baldaniya, D., Jogani, D., Chaudhary, B., Shah, M., & Kshirsagar, A. (2021). Artificial Intelligence: New Age of Transformation in Petroleum Upstream. *Petroleum Research, 7*(1), 106–114.
18. Koroteev, D., & Tekic, Z. (2021). Artificial Intelligence in Oil and Gas Upstream: Trends, Challenges, and Scenarios for the Future. *Energy and AI, 3*, 100041.
19. https://www.digitalrefining.com/article/1000847/manufacturing-execution-systems-for-the-refining-industry#.YtZNK3ZByUk [Online]
20. Al-Fattah, S. M. (2019). Artificial Intelligence Approach for Modeling and Forecasting Oil-Price Volatility. *SPE Reservoir Evaluation & Engineering, 22*(3), 817–826.
21. Ahmad, S., Mishra, S., & Sharma, V. (2023). Green Computing for Sustainable Future Technologies and Its Applications. In Grima, S., Sood, K., & Özen, E. (Eds.), *Contemporary Studies of Risks in Emerging Technology, Part A (Emerald Studies in Finance, Insurance, and Risk Management)* (pp. 241–256). Emerald Publishing Limited, Bingley. https://doi.org/10.1108/978-1-80455-562-020231016
22. Durgalakshmi, K., Anbarasu, P., Karpagam, V., Venkatesh, A., Kannapiran, B., & Sharma, V. (2022). Utilization of Reduced Switch Components with Different Topologies in Multi-Level Inverter for Renewable Energy Applications-A Detailed Review. In *2022 5th International Conference on Contemporary Computing and Informatics (IC3I)*, Uttar Pradesh, India, pp. 913–920. https://doi.org/10.1109/IC3I56241.2022.10073430.
23. Pushpalatha, N., Jabeera, S., Hemalatha, N., Sharma, V., Balusamy, B., & Yuvaraj, R. (2022). A Succinct Summary of the Solar MPPT Utilizing a Diverse Optimizing Compiler. In *2022 5th International Conference on Contemporary Computing and Informatics (IC3I)*, Uttar Pradesh, India, pp. 1177–1181. https://doi.org/10.1109/IC3I56241.2022.10072844.
24. Ali, M. A., Balamurugan, B., Dhanaraj, R. K., & Sharma, V. (2022). IoT and Blockchain Based Smart Agriculture Monitoring and Intelligence Security System. In *2022 3rd International Conference on Computation, Automation and Knowledge Management (ICCAKM)*, Dubai, United Arab Emirates, pp. 1–7. https://doi.org/10.1109/ICCAKM54721.2022.9990243.
25. Banerjee, D., Kukreja, V., Hariharan, S., & Sharma, V. (2023). Fast and Accurate Multi-Classification of Kiwi Fruit Disease in Leaves Using Deep Learning Approach. In *2023 International Conference on Innovative Data Communication Technologies and Application (ICIDCA)*, Uttarakhand, India, pp. 131–137. https://doi.org/10.1109/ICIDCA56705.2023.10099755.
26. Juyal, V., Pandey, N., & Saggar, R. (2016). Impact of Varying Buffer Space for Routing Protocols in Delay Tolerant Networks. In *2016 International Conference on Communication and Signal Processing (ICCSP)*. IEEE.
27. Juyal, V., Pandey, N., & Saggar, R. (2016). A Heuristic Light Weight Security Algorithm for Resource Constrained DTN Routing. In *2016 IEEE International Conference on Computational Intelligence and Computing Research (ICCIC)*. IEEE.
28. Juyal, V., Pandey, N., & Saggar, R. (2017). Performance Comparison of DTN Multicasting Routing Algorithms-Opportunities and Challenges. In *2017 International Conference on Intelligent Sustainable Systems (ICISS)*. IEEE.
29. Balamurugan, B., & Venkata Krishna, P. (2017). Simplified and Efficient Framework for Managing Roles in Cloud-Based Transaction Processing Systems Using Attribute-Based Encryption. *International Journal of Computational Science and Engineering, 14*(2), 135–149.
30. Rizwan, P., et al. (2018). Real-Time Big Data Computing for Internet of Things and Cyber Physical System Aided Medical Devices for Better Healthcare. In *2018 Majan International Conference (MIC)*. IEEE.

31. Maurya, A., & Sharma, V. (2022). Facial Emotion Recognition Using Keras and CNN. In *2022 2nd International Conference on Advance Computing and Innovative Technologies in Engineering (ICACITE)*, pp. 2539–2543. https://doi.org/10.1109/ICACITE53722.2022.9823480.
32. Singh, A., Ali, M. A., Balamurugan, B., & Sharma, V. (2022). Blockchain: Tool for Controlling Ransomware Through Pre-Encryption and Post-Encryption Behavior. In *2022 Fifth International Conference on Computational Intelligence and Communication Technologies (CCICT)*, pp. 584–589. https://doi.org/10.1109/CCiCT56684.2022.00107.
33. Ali, M. A., Balamurugan, B., & Sharma, V. (2022). IoT and Blockchain Based Intelligence Security System for Human Detection using an Improved ACO and Heap Algorithm. In *2022 2nd International Conference on Advance Computing and Innovative Technologies in Engineering (ICACITE)*, pp. 1792–1795. https://doi.org/10.1109/ICACITE53722.2022.9823827.
34. Sharma, V., Balusamy, B., Thomas, J. J., & Atlas, L. G. (Eds.). (2023). *Data Fabric Architectures: Web-Driven Applications*. Walter de Gruyter GmbH & Co KG.
35. Raj, A., Sharma, V., & Shanu, A. K. (2022). Comparative Analysis of Security and Privacy Technique for Federated Learning in IOT Based Devices. In *2022 3rd International Conference on Computation, Automation and Knowledge Management (ICCAKM)*, Dubai, United Arab Emirates, pp. 1–5. https://doi.org/10.1109/ICCAKM54721.2022.9990152.

5 Effects of Contaminated Ocean Water on the Ecosystem

A Review Using Machine Learning

Samiksha Sharma, Aman Verma, Kajal K., and Siva Rama Krishnan Somayaji

CONTENTS

5.1 Introduction 92
5.2 Literature Survey 93
5.3 Architecture 93
5.4 Experimental Setup 101
Data Set: Indian Water Quality Data 101
5.4.1 Two-Class Neural Network Component 103
5.4.2 Decision Forest Algorithm 104
5.4.3 Support Vector Machines 104
5.4.4 Analysis of Results 105
5.5 Conclusion 106
References 106

DOI: 10.1201/9781003348313-5

5.1 INTRODUCTION

The integrated study of ocean health and human health is becoming a growing global priority. According to mounting evidence, the health of the ocean and humans have always been and will continue to be inextricably linked. Our actions toward the oceans will have a significant impact on the entire planet's future, as well as our own health (Figure 5.1).

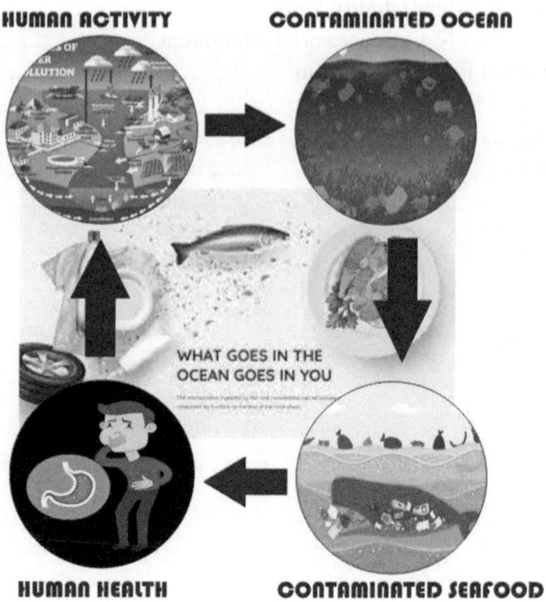

FIGURE 5.1 Dependency of human and ocean health on each other.

The ocean pollutants of which we have the best understanding of human health effects are methylmercury and PCBs. Infants exposed to these pollutants in utero as a result of their mothers' consumption of contaminated seafood can harm developing brains; lower IQ; and increase the risk of autism, attention deficit hyperactivity disorder, and learning disorders in children. Methylmercury exposure in adults increases the risk of cardiovascular disease and dementia. In [1] Manufactured chemicals found in plastic waste, such as phthalates, bisphenol A, flame retardants, and per fluorinated chemicals, can disrupt endocrine signalling, reduce male fertility, harm the nervous system, and increase cancer risk. Harmful algal bloom (HAB)-produced toxins accumulate in fish and shellfish. Toxins can cause severe neurological impairment and death when consumed. HAB toxins can also become airborne and cause respiratory disease.

5.2 LITERATURE SURVEY

See Table 5.1 for a review of the literature.

5.3 ARCHITECTURE

The Internet of Things, or IoT, is a collection of edge- and cloud-based managed and platform services that connect, monitor, and control billions of IoT assets. also includes device and equipment security and operating systems, as well as data and analytics to assist businesses in developing, deploying, and managing IoT applications. The Azure IoT Hub is a cloud-hosted managed service that serves as a central message hub for communication between an IoT application and its attached devices. You can reliably and securely connect millions of devices and their backend solutions. IoT and machine learning uncover insights previously hidden in data, allowing for faster, more automated responses and better decision-making. By utilizing images, videos, and audio, machine learning for IoT can be used to forecast future trends, detect anomalies, and augment intelligence. By analyzing massive amounts of data with sophisticated algorithms, machine learning can help decode the hidden patterns in IoT data. Machine learning inference can supplement or replace manual processes in critical processes with automated systems that use statistically derived actions.

Data for the water data set is collected using water sensors. The various parameters used in this work are explained in the following (also see Figure 5.2).

One of the most important indicators of water quality is *dissolved oxygen* (DO), i.e., the level of free oxygen present in a body of water. Dissolved oxygen sensors for IoT are used to accurately measure the oxygen content of water and transmit the results in real time to a central database, the cloud, or water monitoring platform (Figure 5.3).

The potential of hydrogen (*pH*) of water influences the majority of chemical and biological processes. It is a major environmental factor limiting species distribution in aquatic habitats. Different species thrive in different pH ranges (Figure 5.4).

TABLE 5.1
Summary of Existing Work

S.No.	Article Title	Publication Date	Authors	Scientific Contribution	Reasons for Choosing article
1.	A review of the application of machine learning in water quality evaluation [2]	June 2022	Mengyuan Zhu, Jiawei Wang, Xiao Yang, Yu Zhang, Linyu Zhang, Hongqiang Ren, Bing Wu, Lin Ye	The paper establishes the present shortcomings of the integration of machine learning with water management systems, such as insufficiency in getting enough data with high accuracy due to cost or technical constraints, inability to extend the applications of machine learning in all real water treatment and management systems keeping in mind the complexities involved, and the need for researchers to have a specific set of specialized knowledge.	The paper calls for a need to develop more sophisticated sensors that can more accurately gather data to help improve the accuracy of water quality measurement. New global algorithms and models specially designed to deal with water management and treatment are also needed. And people from different backgrounds should be specialized in this domain, as they can suggest more and better systems for the problem.
2.	IoT based smart water quality monitoring system [3]	November 2021	Varsha Lakshmikantha, Anjitha Hiriyannagowda, Akshay Manjunath, Aruna Patted, Jagadeesh Basavaiah, Audre Arlene Anthony	The article provides an overview of recent research efforts to develop intelligent, energy-efficient, and highly effective water quality monitoring systems that will allow for continuous monitoring and the transmission of alarms and notifications to the appropriate authorities for follow-up action. The created model is inexpensive and easy to use (flexible). Three water samples are examined, and based on the findings, it is possible to determine if the water is fit for consumption.	The article sheds light on recent developments made in the arena of smart water pollution monitoring systems, which worked as a source of direction for us. Also a proposed model that is cost effective and efficient in monitoring quality parameters helped us gain insights on the system.

(Continued)

TABLE 5.1 (CONTINUED)
Summary of Existing Work

S.No.	Article Title	Publication Date	Authors	Scientific Contribution	Reasons for Choosing article
3.	A data-driven method for detecting and diagnosing causes of water quality contamination in a dataset with a high rate of missing values [4]	October 2020	Raymond Houe Ngounaa, Romy Ratolojanahary, Kamal Medjaher, Fabien Dauriac, Mathieu Sebilo, Jean Junca-Bourie	A PHM-based methodology for the identification and diagnosis of water quality problems was put out in this work. This methodology made it possible to derive pertinent conclusions (confirmed by hydrogeology experts) from a heterogeneous data set with a high proportion of missing data. An indirect approach using the European was put out in relation to the diagnostics problem, allowing the transition from an unsupervised problem to a supervised one. The main sources of water pollution – or, more specifically, the sources of water quality disqualification – were finally identified as nitrate and atrazine, and this conclusion was supported by hydrogeology experts participating in this study.	High rates of missing data and measurement invariability cannot be resolved by normal techniques. The reason for choosing this paper was it uses prognostics and health management to identify and treat abnormalities in water quality data sets, taking into consideration the aforementioned problem.

(Continued)

TABLE 5.1 (CONTINUED)
Summary of Existing Work

S.No.	Article Title	Publication Date	Authors	Scientific Contribution	Reasons for Choosing article
4.	Advances in smart environment monitoring systems using IoT and sensors [5]	May 2020	Silvia Liberata Ullo, G. R. Sinha	Genuine environmental difficulties are caused by a number of important elements, including poor air quality, water pollution, and radiation pollution. A healthy society must be maintained in order for the planet to experience sustained growth. Research is aimed at bringing out significant advances in IoT and sensor technologies used for addressing the challenges in smart environment monitoring, so we included some noteworthy research studies and contributions from various sources. We found some interesting literature on specific areas of research addressing some challenges of environmental factors such as water pollution, air quality, and contamination in ocean water.	A detailed and critical assessment of research studies on various environmental monitoring systems used for a variety of reasons has been provided in this publication. Significant recommendations were provided by the review's analysis and discussion. It has been realized that there is a need for in -depth study on deep learning, addressing huge data and noisy data difficulties, and developing a framework of reliable classification algorithms. The monitoring of air and water quality that can address environmental issues has received the majority of our attention.

(Continued)

TABLE 5.1 (CONTINUED)
Summary of Existing Work

S.No.	Article Title	Publication Date	Authors	Scientific Contribution	Reasons for Choosing article
5.	Machine learning approaches for anomaly detection of water quality on a real-world data set [6]	February 2019	Fitore Muharemi, Doina Logofătu, Florin Leon	Accurately detecting changes in water quality is a crucial duty that water companies must carry out. Water providers are required to provide safe drinking water. Intelligent sensors that continuously monitor data in a variety of locations are now available. The data collected by the sensors are usually significant, but they are occasionally misinterpreted, making it difficult to determine whether an event has occurred. Various methods for detecting changes or anomalies in real-time data on water quality are described in this paper, and a solution is proposed. Water quality data is analyzed using models such as support vector machines, logistic regression, and linear discriminant analysis.	The reason for selecting this work is that it presents methods for detecting events in water quality data. This work is a case study, and the goal is to find the best model for detecting anomalies in water quality systems. Data sets have long been analyzed using statistical algorithms, and machine learning algorithms are well known for performing well on these data sets.
6.	Water quality monitoring system using IoT [7]	February 2018	Nageswara Rao Moparthi, Ch. Mukesh, P. Vidya Sagar	Water pollution has become an increasingly serious problem in recent years. This IoT-based water quality monitoring system's goals are to determine the water's quality, including how pH, turbidity, and other measurements vary; and to send messages to the appropriate authorities so that the necessary actions can be taken while keeping human and marine life, health, and disease in mind.	The purpose of referring to this paper is to learn about the various parameters that contribute to water contamination and how they can be improved to better the lives of aquatic animals and humans through the real-time use of IoT sensors and machine learning algorithms.

(Continued)

TABLE 5.1 (CONTINUED)
Summary of Existing Work

S.No.	Article Title	Publication Date	Authors	Scientific Contribution	Reasons for Choosing article
7.	Beyond the ocean: Contamination of freshwater ecosystems with (micro-) plastic particles	March 2015	Rachid Dris, Hannes Imhof, Wilfried Sanchez, Johnny Gasperi, Francois Galgani, Bruno Tassin, Christian Laforsch	According to some research, microplastics have been discovered in freshwater environments, and the contamination there is just as bad as it is in the oceans. Microplastics have been discovered in sediments (primarily lake shores, but also riverbanks) and water samples taken from continental waterways.	This paper is considered because it presents contamination not only in ocean water, but also in sediments and continental waterways. By analyzing these samples, warnings for the future can be generated.

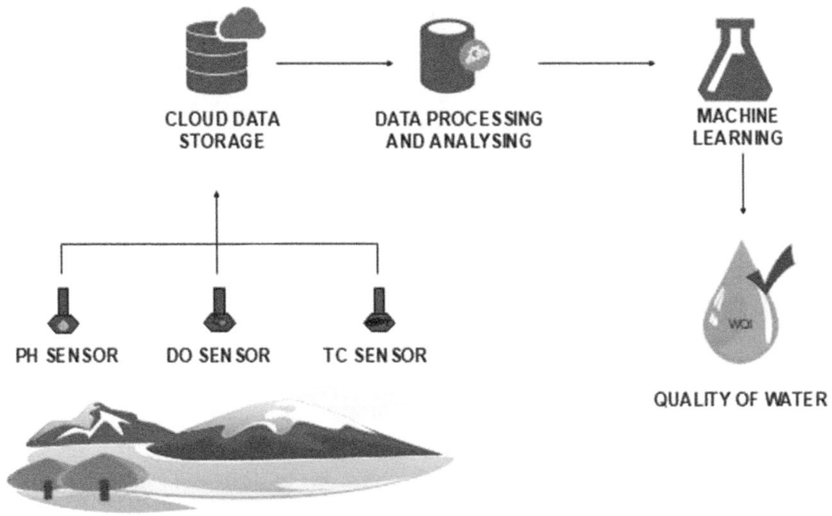

FIGURE 5.2 Proposed architecture.

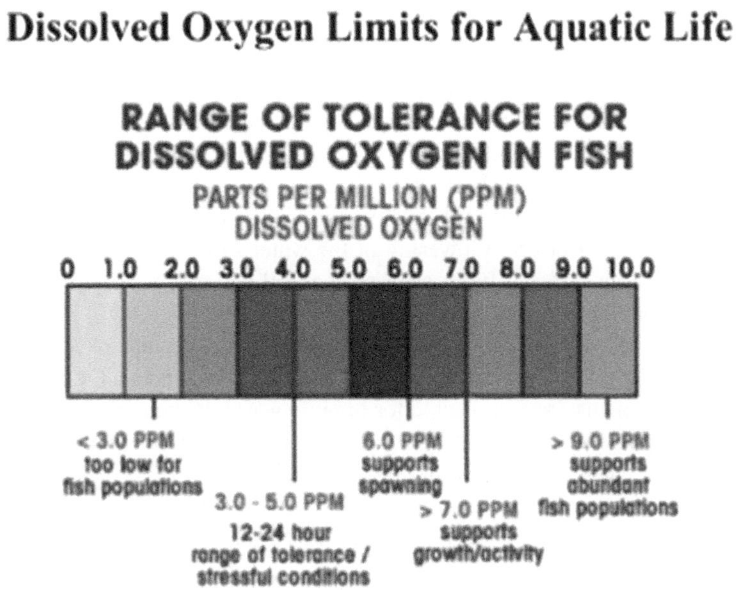

FIGURE 5.3 Dissolved oxygen limits for aquatic life.

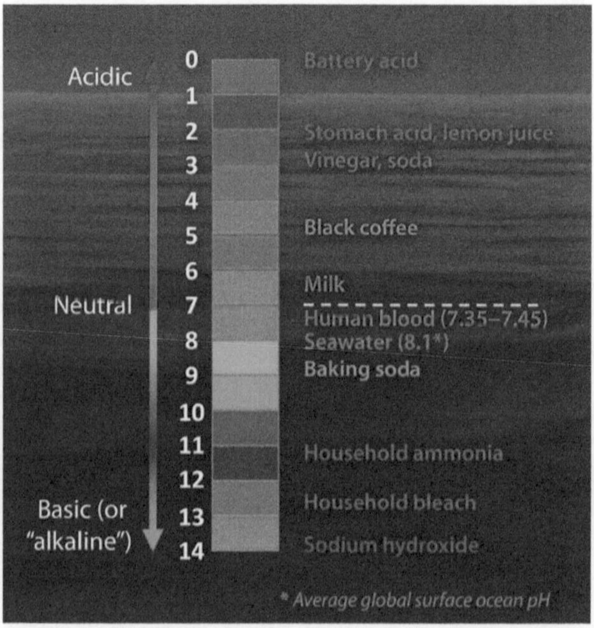

FIGURE 5.4 pH ranges of surface ocean water.

According to the National Oceanic and Atmospheric Administration (NOAA), the pH of surface ocean waters has dropped by 0.1 since the beginning of the industrial revolution. This represents a roughly 30% increase in acidity. If current levels of emissions continue, ocean acidification is expected to increase by 144% by 2100.

Ocean acidification is caused by carbon dioxide gas in the atmosphere dissolving into the ocean. This leads to a lowering of the water's pH, making the ocean more acidic. The burning of fossil fuels, such as coal, oil and gas, for human industry is one of the major causes of ocean acidification.

Biological oxygen demand (BOD) is the amount of molecular oxygen required, in milligrams, to convert organic molecules in a litre of water to CO_2. Bacteria and other microorganisms are responsible for organic waste decomposition. When dead plants, leaves, grass clippings, manure, sewage, or even food waste are present in a water supply, bacteria begin the process of breaking down this waste. When this occurs, aerobic bacteria consume a large portion of the available dissolved oxygen, depriving other aquatic organisms of oxygen. A common metric for water pollution is the BOD level.

The most basic definition of *salinity* is the concentration of dissolved salts in water. Seawater contains salts other than sodium chloride (NaCl), such as calcium, magnesium, and potassium. The density of ocean water can be affected by salinity: water with a higher salinity is denser and heavier, and will sink beneath less saline, warmer water. This can have an impact on marine life, which may need to regulate their saltwater intake.

Large amounts of faecal *coliform* are released in farm animal waste and can be washed into streams by rain or irrigation runoff. When waste from dogs, cats, raccoons, and humans is carried into storm drains, creeks, and lakes during storms, it contributes to faecal coliform contamination. Faecal coliform can also enter streams through illegal or leaky sanitary sewer connections, faulty septic tanks, and inefficient wastewater treatment plants. These bacteria suggest the presence of pathogens or disease-carrying bacteria, such as *E. coli*. Bacteria and pathogens are primarily a concern for human health. When bacteria decompose, they can deplete the oxygen in the water and act as a fertilizer for algae and other aquatic plants, causing an overgrowth.

By using the aforementioned parameters, the Water Quality Index (WQI) is a unique and effective rating technique for determining water quality. However, most of the indices are not applicable to all water types because they are dependent on core physicochemical water parameters that can make them biased and sensitive to specific attributes such as:

1. Data sampling time, location, and frequency.
2. Parameter number, variety, and weight allocation.

As a result, these indices must be evaluated to remove uncertainties that make them unpredictable and may lead to manipulation of water quality classes.

5.4 EXPERIMENTAL SETUP

Data Set: Indian Water Quality Data

Upload the data set in the Machine Learning Studio (Figure 5.5).

Create a new experiment (Figure 5.6).

FIGURE 5.5 Uploaded data set.

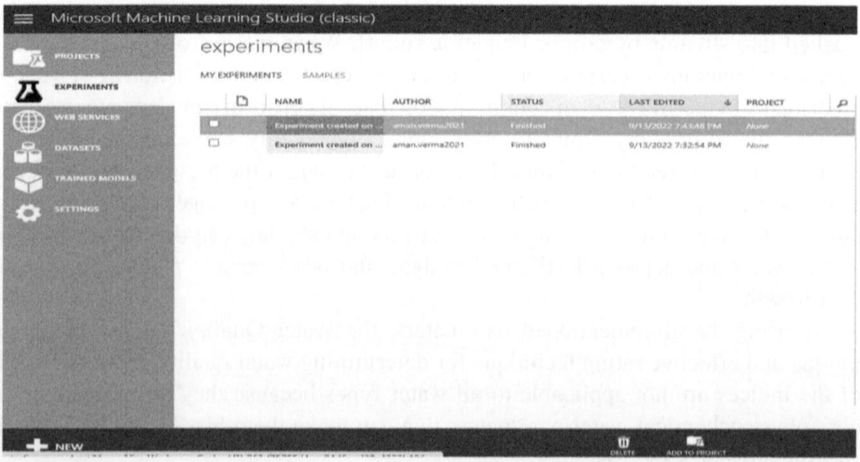

FIGURE 5.6 Experiment created.

Generate a pipeline by drag your data set to the experiment page (Figure 5.7).

First, non-contributing attributes will be removed, and null values will be filled with the mean. When the data size is small, this is a reasonable approach (Figure 5.8).

Data is then fed into specific machine learning algorithms to produce a non-harmful mean quality value. Daily measured values can be regulated based on that value, and when they are exceeded, a corresponding warning can be issued, resolving the issue.

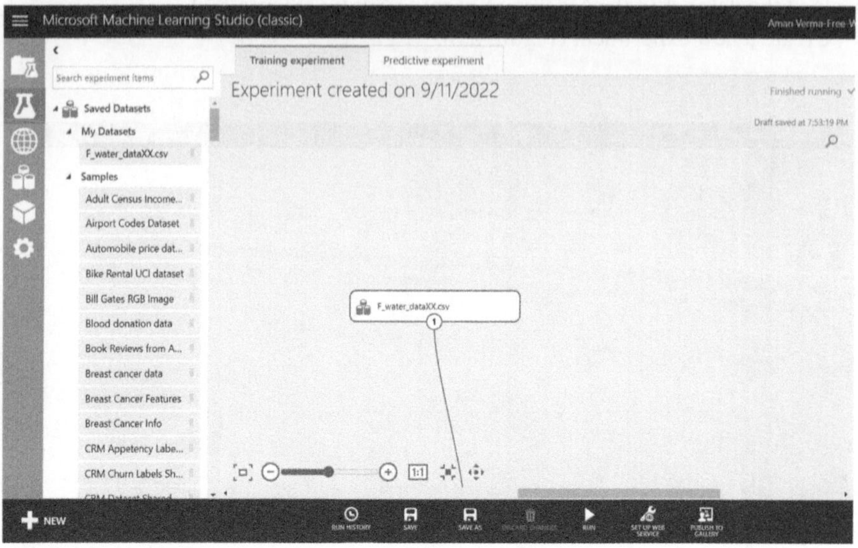

FIGURE 5.7 Drag data set.

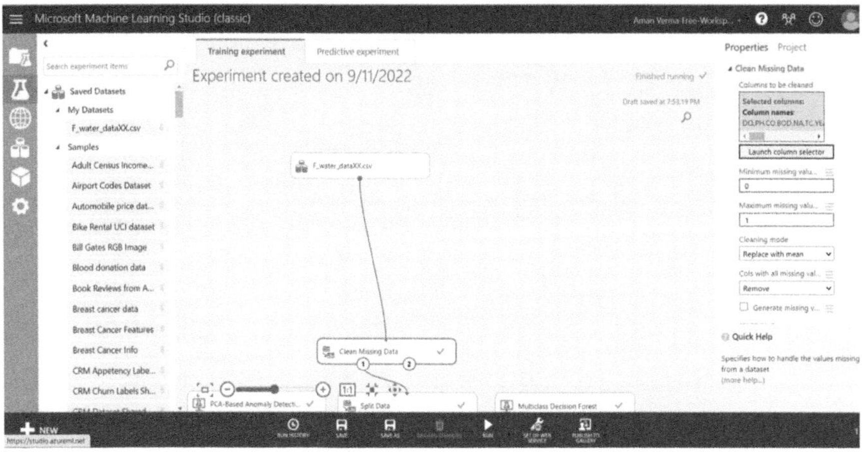

FIGURE 5.8 Training experiment.

5.4.1 Two-Class Neural Network Component

Create a neural network model based on green computing such as in [8] with this component that can forecast a target using only two values. The supervised learning technique of classifying objects using neural networks requires a tagged data set with a label column (Figure 5.9 and Figure 5.10). This neural network model, for example, could be used to predict binary outcomes, such as whether a patient has a specific disease or whether a machine will fail within a specific time frame. [9] suggests the implementation of solar poweredbased equipments for the sustainable and renewable energy source.

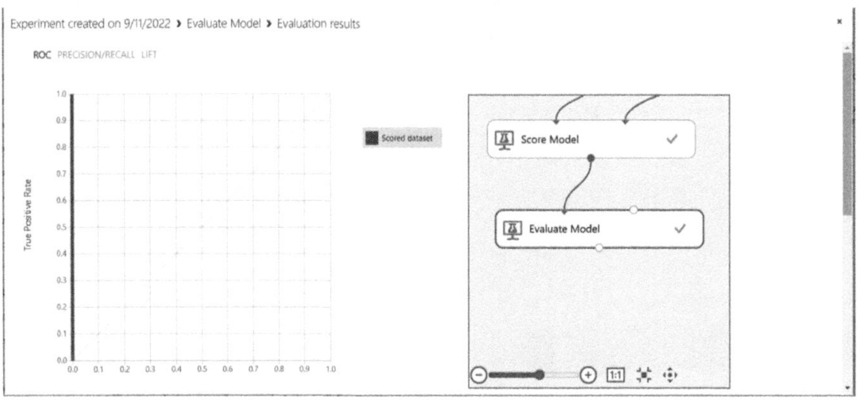

FIGURE 5.9 Evaluate model.

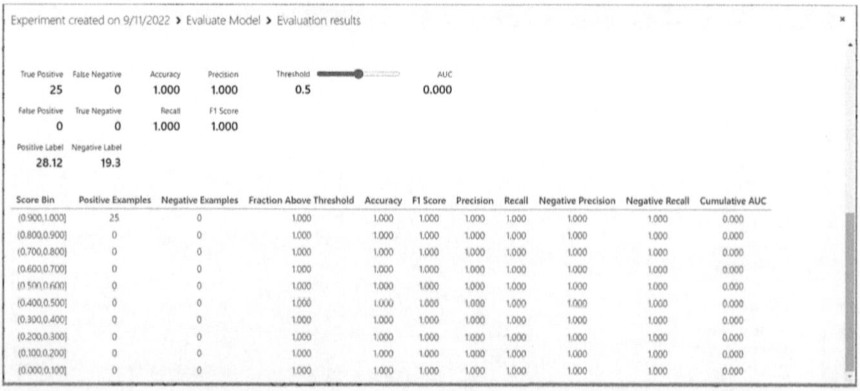

Score Bin	Positive Examples	Negative Examples	Fraction Above Threshold	Accuracy	F1 Score	Precision	Recall	Negative Precision	Negative Recall	Cumulative AUC
(0.900,1.000]	25	0	1.000	1.000	1.000	1.000	1.000	1.000	1.000	0.000
(0.800,0.900]	0	0	1.000	1.000	1.000	1.000	1.000	1.000	1.000	0.000
(0.700,0.800]	0	0	1.000	1.000	1.000	1.000	1.000	1.000	1.000	0.000
(0.600,0.700]	0	0	1.000	1.000	1.000	1.000	1.000	1.000	1.000	0.000
(0.500,0.600]	0	0	1.000	1.000	1.000	1.000	1.000	1.000	1.000	0.000
(0.400,0.500]	0	0	1.000	1.000	1.000	1.000	1.000	1.000	1.000	0.000
(0.300,0.400]	0	0	1.000	1.000	1.000	1.000	1.000	1.000	1.000	0.000
(0.200,0.300]	0	0	1.000	1.000	1.000	1.000	1.000	1.000	1.000	0.000
(0.100,0.200]	0	0	1.000	1.000	1.000	1.000	1.000	1.000	1.000	0.000
(0.000,0.100]	0	0	1.000	1.000	1.000	1.000	1.000	1.000	1.000	0.000

FIGURE 5.10 Evaluation results.

5.4.2 Decision Forest Algorithm

The decision forest algorithm is an ensemble learning classification technique. As part of the method, multiple decision trees are built, and the most popular output class is chosen by vote. During voting, each tree in a classification decision forest produces a non-normalized frequency histogram of labels, which is a type of aggregation. During the aggregation phase, these histograms are added and the result is normalized to produce the "probabilities" for each label. Trees with high prediction confidence are given more weight in the ensemble's final selection. In general, decision trees are non-parametric models that can handle data with varying distributions. A series of simple tests for each class raises the levels of each tree until a leaf node (decision) is reached (Figure 5.11).

5.4.3 Support Vector Machines

Support vector machines (SVMs) are a type of supervised learning technique of Machine learning algorithms and implementation of solar powered equipments [10]

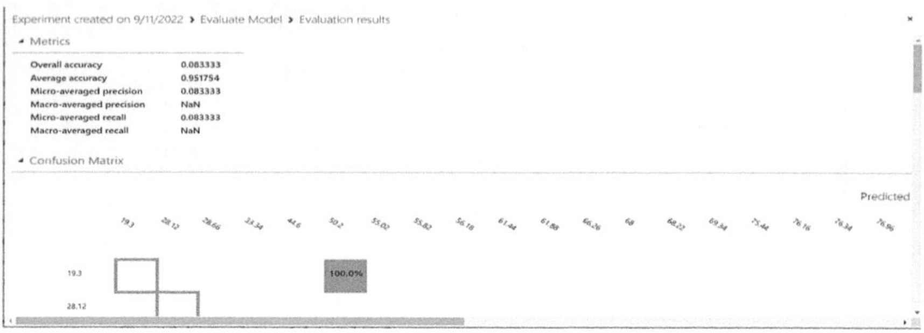

Metric	Value
Overall accuracy	0.083333
Average accuracy	0.951754
Micro-averaged precision	0.083333
Macro-averaged precision	NaN
Micro-averaged recall	0.083333
Macro-averaged recall	NaN

FIGURE 5.11 Evaluation results.

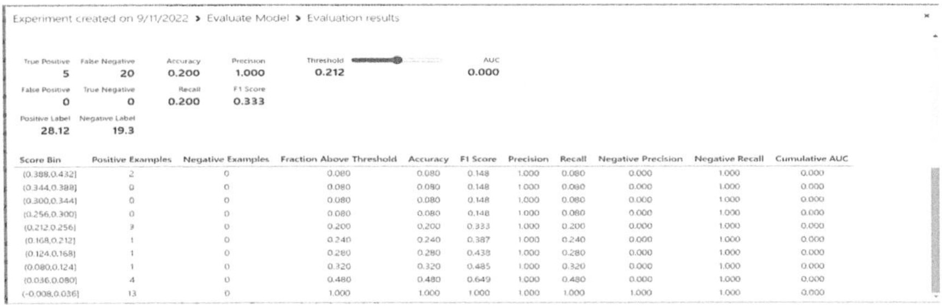

FIGURE 5.12 Experiment results.

that has received a lot of attention. This solution is appropriate for two-outcome prediction based on categorical or continuous data. Once the parameters of the model have been determined, the model should be trained using the training components and a tagged data set with a label or result column (Figure 5.12).

Following model definition, train the model by passing a tagged data set and the model as inputs to Train Model. After training, the model can be used to forecast values for fresh inputs (Figure 5.13).

5.4.4 Analysis of Results

As can be seen from Table 5.2, the accuracy for the two-class neural network and decision forest algorithm is above 95%, which is significantly good to help determine the water quality level and thus, save people from getting affected by contaminated water and the diseases that it brings. Support vector machines have relatively very low accuracy, which provides the scope for improvement and a topic for further studies.

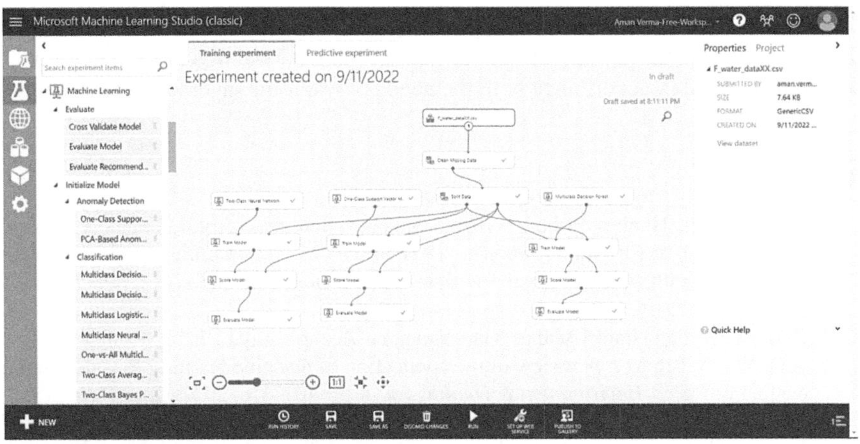

FIGURE 5.13 Model ready to forecast values.

TABLE 5.2
Summary of Results

Model	Accuracy	Precision
Two-class neural network	1.000	1.000
Support vector machines	0.200	1.000
Decision forest algorithm	0.951754	0.083333

The precision values for the two-class neural network and support vector machine is 100%, which is the number of true positives per predicted positives. However, the decision forest algorithm has much less precision, due to decision forests being too sensitive to change, thus resulting in minimal correlation, but it still has an accuracy of over 95%. Having a precision value of 1.0 means that the model produces no false positives, so two-class neural networks and support vector machines can be used in cases where false negatives matter less than false positives.

5.5 CONCLUSION

Several machine learning algorithms were used to predict the quality of water and it is found that the accuracy for the two-class neural network and decision forest algorithm is above 95%, which is significantly good to help determine the water quality level and thereby prevent people from contracting diseases and being exposed to contaminated water. Machine learning has been widely used as a powerful tool to solve problems in the water environment because it can be applied to predict water quality. These methods have some implementation challenges, therefore in the future, several parameters should be taken into account, such as the development and use of more sophisticated sensors, to collect data that is accurate enough to support the application of machine learning techniques. Improvements should be made to the algorithms' sustainability and reliability, and new global algorithms and models should be created in accordance with the needs of water management and treatment.

REFERENCES

1. Dris, Rachid, Hannes Imhof, Wilfried Sanchez, Johnny Gasperi, François Galgani, Bruno Tassin, & Christian Laforsch, "Beyond the ocean: Contamination of freshwater ecosystems with (micro-) plastic particles," *Environmental Chemistry*, vol. 12, no. 5, pp. 539–550, 2015.
2. Zhu, Mengyuan, Jiawei Wang, Xiao Yang, Yu Zhang, Linyu Zhang, Hongqiang Ren, Bing Wu, & Lin Ye, "A review of the application of machine learning in water quality evaluation," *Eco-Environment & Health*, vol. 1, pp. 107–116, 2022.
3. Lakshmikantha, Varsha, Anjitha Hiriyannagowda, Akshay Manjunath, Aruna Patted, Jagadeesh Basavaiah, & Audre Arlene Anthony, "IoT based smart water quality monitoring system," *Global Transitions Proceedings*, vol. 2, no. 2, pp. 181–186, 2021.

4. Ngouna, Raymond Houé, Romy Ratolojanahary, Kamal Medjaher, Fabien Dauriac, Mathieu Sebilo, & Jean Junca-Bourié, "A data-driven method for detecting and diagnosing causes of water quality contamination in a dataset with a high rate of missing values," *Engineering Applications of Artificial Intelligence*, vol. 95, 103822, 2020.
5. Ullo, Silvia Liberata, & G. R. Sinha, "Advances in smart environment monitoring systems," *Internet of Things*, vol. 20, no. 11, 3113, 2020.
6. Muharemi, Fitore, Doina Logofătu, & Florin Leon, "Machine learning approaches for anomaly detection of water quality on a real-world data set," *Journal of Information and Telecommunication*, vol. 3, no. 3, pp. 294–307, 2019.
7. Moparthi, Nageswara Rao, Ch. Mukesh, & P. Vidya Sagar, "Water quality monitoring system using IOT," *IEEE*, vol. 4, pp. 1–5, 2018.
8. Pushpalatha, N., S. Jabeera, N. Hemalatha, V. Sharma, B. Balusamy, & R. Yuvaraj, "A succinct summary of the solar MPPT utilizing a diverse optimizing compiler," *2022 5th International Conference on Contemporary Computing and Informatics (IC3I)*, Uttar Pradesh, India, pp. 1177–1181, 2022, doi: 10.1109/IC3I56241.2022.10072844.
9. Durgalakshmi, K., P. Anbarasu, V. Karpagam, A. Venkatesh, B. Kannapiran, & V. Sharma, "Utilization of Reduced Switch Components with Different Topologies in Multi-Level Inverter for Renewable Energy Applications-A Detailed Review," *2022 5th International Conference on Contemporary Computing and Informatics (IC3I)*, Uttar Pradesh, India, pp. 913–920, 2022, doi: 10.1109/IC3I56241.2022.10073430.
10. Ahmad, S., S. Mishra, & V. Sharma, "Green Computing for Sustainable Future Technologies and Its Applications," in Grima, S., Sood, K., & Özen, E. (Eds.), *Contemporary Studies of Risks in Emerging Technology, Part A (Emerald Studies in Finance, Insurance, and Risk Management)*, Emerald Publishing Limited, Bingley, pp. 241–256, 2023, doi: 10.1108/978-1-80455-562-020231016.

6 Trends and Challenges in Green Computing

Priti Rai and Yogita Rawat

CONTENTS

6.1 Introduction ... 109
6.2 Need of the Study ... 110
6.3 Objective of the Study ... 111
6.4 Review of Literature ... 111
6.5 Green Computing and Its Necessity ... 112
6.6 History of Green Computing ... 114
6.7 Benefits of Green Computing ... 115
6.8 Advantages of Green Computing ... 117
6.9 Limitations in Green Computing Adoption ... 117
6.10 Disadvantages of Green Computing ... 118
6.11 Implementation Strategies of Green Computing ... 118
6.11.1 Procedure for Implementing Green Computing ... 118
6.11.2 Green Computing Solutions ... 119
6.11.3 Green Computing Practices ... 121
6.12 Implementing Strategies for Green Computing ... 124
6.12.1 Industry and Government Schemes in Favour of Green Computing ... 126
6.12.2 Sustainable Development with Green Computing in Post-COVID Era ... 128
6.13 Future of Green Computing ... 129
6.14 Conclusion ... 131
References ... 132

6.1 INTRODUCTION

Green computing, popularly green IT or sustainable IT, explains the design, manufacture, utility, and disposing of a computer system, its technical components, and external devices in a way that minimizes the negative environmental impact of the equipment. Examples include lowering the energy used by businesses, data centres, and end users, as well as greenhouse gas (GHG) emissions (Figure 6.1).

On the lines of the changing psychological perspective, technology has evolved greatly, but it requires an enormous amount of energy to produce, store, pack, and deliver them. Computers have typically been produced using toxic elements like

DOI: 10.1201/9781003348313-6

FIGURE 6.1 Green computing.

cadmium, lead, and mercury. Two-fifths of the lead found in landfills is thought to come from computers and other gadgets. An eco-friendly computer is something that people are looking for in order to combat the growing threat of pollution that exists on a global scale as a result of the increased consumption of electronic devices, especially computers systems. Advanced processors require more power, which results in increased heat production from their waste. As a result, it's crucial to cool and aerate them, especially in server farms, between the HVAC (Heating Ventilation Air Conditioning) and computers components.

The most important question is why green computing is required. A computer left running for a long duration of time will result in high electronic costs and excess emission of atmospheric carbon dioxide. But nowadays, a growing number of computers are generating extra greenhouse gases every year that contribute to global warming. Due to this, an eco-friendly approach to computer usage is crucial for software developers. Learning about green computing is therefore essential. It highlights the effects of a computer system on the environment over its lifespan. Making a sustainable computer device last longer helps in decreasing the impact of its usage. As the disposing of computer waste, which plays a crucial part in a computer's life cycle, is kept on hold, it should be considered to use a computer for as long as possible to reduce its ill impact on the environment. The environmental impact of manufacturing the latest computer device from scratch is greater than that of creating a new RAM module to replace one in computing hardware. Making major advances in green computing technology is made possible by computer virtualization, which helps in utilizing multiple computers on a single hardware. This might be used to create the theoretically perfect green computer, which does not physically exist.

6.2 NEED OF THE STUDY

As technology plays an increasing part in our lives, there is so much pollution released from the computer and other devices, and it is crucial to lessen the factors that contribute to pollution. Green computing seeks to minimize hazardous materials, increase power saving throughout the device's lifespan, and encourage the recycling or disposing of outdated devices and managing waste. Incorporating sustainable IT techniques and strategies in daily life can quickly lower bills, decrease e-waste, and save on energy.

6.3 OBJECTIVE OF THE STUDY

The objective of the study is to examine the requirement for green computing and its implementation strategies in today's digital world to prevent electronic hazards with proper implementation of reusability, recyclability, and biodegradability.

6.4 REVIEW OF LITERATURE

The article "Green Computing: An Era of Energy Saving Computing of Cloud Resources" suggested that green computing can help reduce the excessive consumption of energy by implying an ON/OFF approach. According to research presented in the paper, the most difficult task currently in the green cloud environment is to transform energy sustainability into energy efficiency. The suggested method of turning on only the resources that are necessary and turning off the rest in order to save energy while in use transforms cloud computing into green computing. Due to lower electricity costs and less CO_2 emissions, this decrease in energy use had an impact on cloud computing service costs [1].

The study "Green Computing – Need and Implementation" suggested that computers are now a necessity for individuals and companies. No single person or company is now able to function in the absence of a computer, but they must be familiar with the negative effects of computer use, as well as of computer manufacturing and disposal, and the steps that should be taken to lessen these effects and protect the environment. The authors suggested that without proper green computing implementation, the environment would suffer various pollutions and a drastic effect on the environment and as well as humans [2].

In their article titled "Green Computing: An Overview with Reference to India", Chowdhury et al. suggested that implementing green computing strategies makes sense from a business standpoint as well as an ethical or moral one [3]. A green computing strategy can be implemented for many business benefits, including cost savings, adaptability, disaster recovery plan, incident management, and media relations. Every single person can contribute towards the goal of decreasing e-waste and promoting power savings by implementing green IT habits. The computing sector seems to be more knowledgeable and ready than practically any other industry when it comes to embracing and adjusting to fast change. Thus, it is reasonable to believe that within a few years, most computers will be using significantly less energy than they do so unintentionally right now [4].

The article titled "Environment-Friendly Green Computing in India" suggested that when it involves adapting to rapid change, the computing industry, unlike any other industry, is much more capable and prepared. 'the majority of PCs wind up in landfills after only a few years of use, primarily the ones used in businesses [5]. This suggests that there is already a general determination to make adjustments to and routinely invest in new computer equipment. Therefore, compared to the decades it took to create more power-saving transportation on the market, it should only take a few years for the bulk of computers to take significantly less electricity than they currently do. Every action uses energy, so the best solution isn't always to buy a new, more efficient computer. Therefore, it can be concluded with confidence that all interested parties must cooperate in order to maintain a safe and clean environment for our coming generations by adopting green computing strategies [6].

In their article "Environmental Sustainability and Green Policies of ICT Industry", Bansal et al. stated that the GHG emission guidelines put in place by the government to control emissions of GHG from various sectors have helped in safeguarding the environment from e-hazards. Companies in the private sector are adjusting their policies to adhere to governmental regulations [7]. According to an analysis of the policies set forth by the governments of various nations to reduce energy consumption, governments are introducing a number of initiatives to encourage private sector businesses to do the same and also purchase renewable energy from local governments to minimize the effects. Governments have established guidelines and regulations for data centres to track their energy efficiency. To increase data centre energy efficiency, both the government and private companies are migrating toward virtualization. In order to improve computing in a manner that has as little influence as possible on the environment, research on green computing is an ongoing effort that necessitates collaboration of ideas and research. The research goal is to analyze the research done by IT companies on green computing, define technologies often used to make computing more efficient, and provide a detailed analysis of the best technologies used in computing to decrease its environmental consequences. This research will be presented in upcoming publications [8].

In this chapter, easy-to-implement strategies are discussed that can help in implementing green computing in daily life to reduce electronic hazards.

6.5 GREEN COMPUTING AND ITS NECESSITY

Computers are now a necessity for every person (Figure 6.2). The use of IT devices saves a lot of time and makes our lives comfortable, but it has also drastically increased power consumption and increased heat production [9]. Greenhouse gases like carbon dioxide (CO_2), containing a number of detrimental effects on our environment and eco-resources, are released when power consumption and heat production increase. This is a result of our ignorance towards the drastic effects of using computers on the environment. Data centres and personal computers consume enormous energy and operate with inadequate cooling systems where a polluted environment is the end result.

The term "green computing"," also known as "eco-ICT", refers to information and communications technology systems that are created in such a way as to decrease the adverse effects of human activity on the environment [10]. It was first used in the *Official Journal of the French Republic* on July 12, 2009. Modern operating

FIGURE 6.2 Green computing should be applicable on all electronic devices.

systems (OS) use an advanced configuration and power interface (ACPI), promote power conservation, and enable users to program the monitors and storage devices to turn off after a specified period of inactivity. Virtual computers can be operated on a single hardware server with the help of virtualization technologies.

Nowadays, most people are aware of the value of green computing technology in protecting the universe from pollution and other environmental dangers. Over the past century, the world has seen dramatic changes in climate, water, and air pollution conditions [11]. The difficulties that the universe faces include severe droughts, ocean acidification, increasing sea levels, increased groundwater depletion, river pollution, and an overall increase in the Earth's temperature. The quick spread of illnesses, viruses, macroparasites, and the loss of animal species are all directly caused by environmental changes. Most of these modifications are irreversible. Nations, governments, communities, and individuals must act rapidly to rescue the Earth [12]. Children and young people have taken the lead in global efforts to reduce air pollution, water pollution, and climate change over the past ten years. Green computing needs to be implemented immediately in order to save our environment. Green computing may be achieved by safeguarding the environment from harmful substances and their effects that are created by computers and connected equipment. Following are the four approaches that are used to promote green computing throughout our day-to-day lives (Figure 6.3):

1. **Green design** – It is the practice of developing products and their features that respect the environment, for example, environmentally friendly printers, computers, and data centres.
2. **Green manufacturing** – It refers to the identification and creation of unique goods that lessen or deal with the production of waste materials in manufacturing.

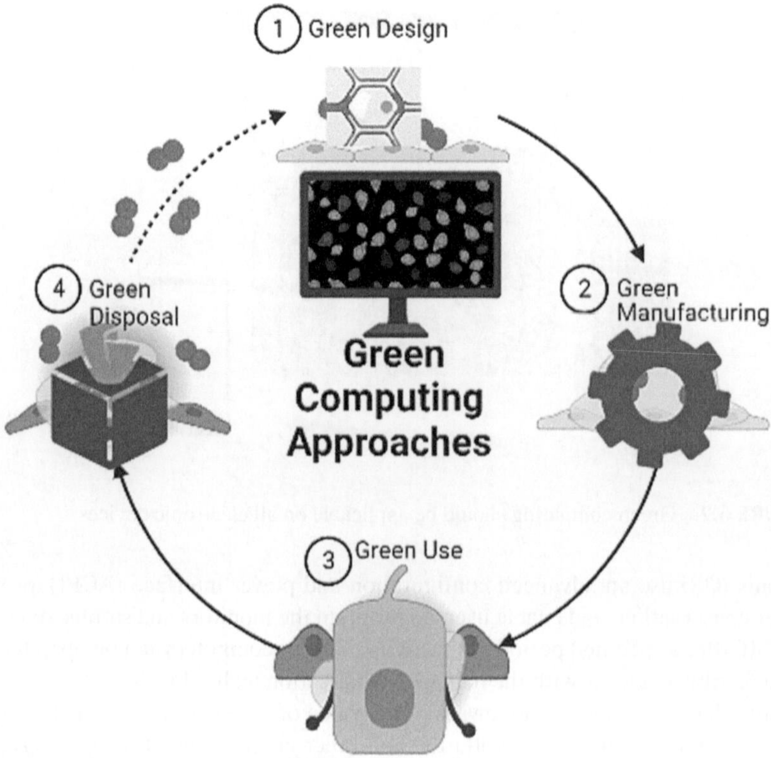

FIGURE 6.3 Green computing approaches.

3. **Green use** – It is the utilization of IT devices in an environmentally sustainable manner that minimizes the use of harmful materials.
4. **Green disposal** – It refers to utilizing already-existing materials or recycling e-waste with minimal or no environmental impact.

6.6 HISTORY OF GREEN COMPUTING

One of the first instances of the green computing movement was the 1992 introduction of the Energy Star programme [13]. Energy Star certification was voluntarily awarded to computing devices that were successful in decreasing energy usage while maximizing efficiency. Computer monitors, television sets, and other products that regulate temperature, such as air conditioners, refrigerators, and other similar products, all qualify for the Energy Star label. The sleep mode feature of computer displays, which switches a user's electrical devices to sleep mode after a certain length of time has passed with no user activity detected, was one of the first applications of green computing. In addition to thin client technologies, energy cost accounting, virtualization strategies, e-waste, etc., the notion of green computing has grown.

Tactical instrumentalists are recently the most well-known green computing organizations. Green computing principles are applied and used by this group more for cost-cutting than for environmental protection [14]. The idea of green computing developed naturally as a result of the pressure placed on businesses to make the most of their resource base to successfully compete in the market. Rather than being influenced by politics, this movement was primarily driven by economic sentiments.

The government has helped advance the idea of green computing for some time. For instance, the US Environmental Protection Agency (EPA) launched the Energy Star programme in the 1990s to encourage energy-saving behaviours (see earlier). The EPA continues to play a crucial role by providing customers with strategies that are both energy and cost efficient. In order to save money, the EPA revamped the programme in 2006 "with the goal of saving U.S. consumers and companies more than \$1.8 billion in energy expenses over the next five years, acting as the basis for regular policy changes and progress in the field of green computing [15].

6.7 BENEFITS OF GREEN COMPUTING

There are many good reasons, but it makes more sense to drive innovation and computation with lower costs and less environmental impact. Here are the benefits of green computing (also see Figure 6.4):

1. **Resource usage** – Green computing aids in the effective and efficient use of resources, such as infrastructures offices, data centres, computers, lighting, and electricity [16]. Balancing resources with actual demand, it helps to reduce resource waste. For instance, systems with lower workloads can be combined with systems with a fair workload with the same energy and resources.
2. **Saving money** – Because natural resources are finite, electricity prices are currently rising daily. Utilizing green computing techniques can lower operating expenses for businesses in areas like energy use, management, maintenance, disposal, and recycling. For instance, it is preferable to make use of the existing IT assets rather than purchasing or adding new ones to the existing infrastructure.
3. **Sustainability of the environment** – Green IT ensures biological system diversity and productivity by minimizing the adverse effects of manufacturing processes through the green approach of design, manufacturing, use, and disposal practices. Hazardous materials have less of an effect on the environment as a result.
4. **Improved social and corporate image** – Businesses can use environmental sustainability as a corporate social responsibility (CSR) strategy to increase customer loyalty, brand recognition, and competitive advantage. By taking part in activities that support green practices, employees are more effectively engaged and may grow to be more loyal to the business.

FIGURE 6.4 Green computing benefits.

5. **Meeting compliance and regulatory requirements** – By meeting compliance and regulatory requirements as well as the emergence of new laws and regulations in specific regions, green IT aids businesses in enhancing their brand awareness and environmental responsibility.
6. **Risk management** – For any business, risk management is their top priority. Practices against the environment can destroy a brand and reputation that took decades to establish in a matter of hours. Environmentally friendly business practices can shield organizations from legislative amendments, escalating fuel costs, negative publicity, and shifting consumer demands.

In order to get the full benefits of green computing, organizations, businesses, and governments should all together implement green computing strategies to protect the environment against electronic hazards [17].

6.8 ADVANTAGES OF GREEN COMPUTING

Following are the advantages of using green computing,

1. Green computing methods decrease energy use, which lowers carbon dioxide emissions since fewer fossil fuels are utilized in transportation and power plants.
2. When resources are saved, it takes less energy to produce, utilize, and dispose of items.
3. Energy and resource conservation are cost effective.
4. The green computing movement includes even modifying governmental regulations to encourage recycling and lower energy use by consumers and businesses.
5. Minimize the danger posed by substances found in computers that are known to cause cancer, immune system responses, and neurological damage.

6.9 LIMITATIONS IN GREEN COMPUTING ADOPTION

More and more businesses are investing in green computing as computing becomes more widespread and expensive, but there are still some issues that need to be resolved and obstacles overcome that prevent companies from implementing green ideas into their business processes [18]. The biggest obstacle preventing businesses from implementing green IT is cost. Figure 6.5 displays the main obstacles that survey respondents identified.

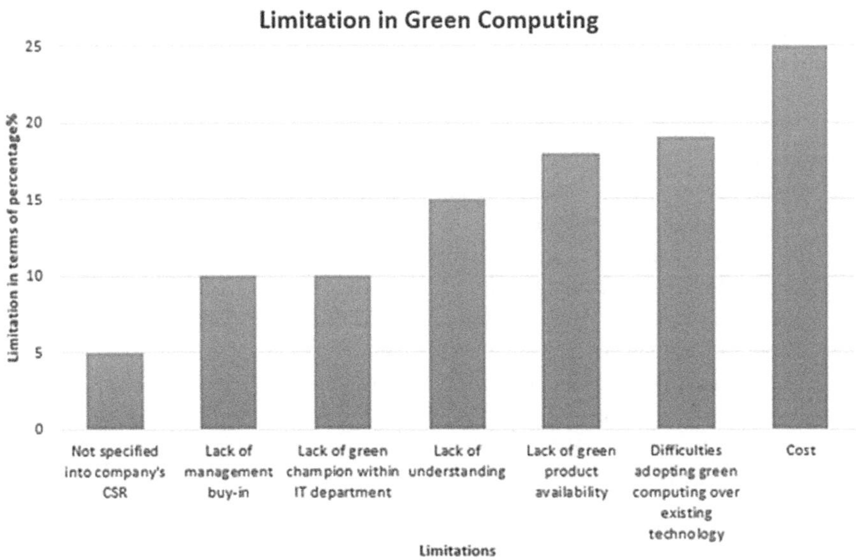

FIGURE 6.5 Limitations for adopting green computing.

These obstacles may exist for businesses adopting green IT, but they must be overcome in order to help people move from concern to action. Going green benefits not only businesses but also entrepreneurs, people, and the general public [19]. To put into practice green IT practices, everyone should work both independently and collaboratively.

6.10 DISADVANTAGES OF GREEN COMPUTING

Green computing might end up being quite expensive. When a computer is produced, there will typically be some additional cost because the computer may have undergone a green manufacturing process in the first place. Since green computing necessitates a substantial amount of brand-new technology, you might discover that the cost of your new green computer is higher than average. The fact that MacBooks are currently the best green computers is a prime illustration. These computers are among the most expensive ones on the market [20].

6.11 IMPLEMENTATION STRATEGIES OF GREEN COMPUTING

6.11.1 Procedure for Implementing Green Computing

It's one thing to acknowledge the necessity of adopting green IT; it's another to actually do it. Numerous studies have been done on how to make IT departments more environmentally friendly. As a result, the IT sector as a whole is busting at the seams with data on green computing, many of which may even conflict with other environmental initiatives. Green computing modification lessens computers negative impact on the environment by taking into account the skills and jobs of computer users and organizations using computers and their components [21]. The following are some actions considered necessary:

1. Virtualization – All servers are divided up into a variety of virtual machines that carry out lively functions in virtualization, which is an amazing electrical device for users and environmentally friendly processing. Organizations can increase the rate at which servers are used by utilizing these fields.
2. Extra-efficient processors are a crucial component of saving energy, and Sun Microsystems, Intel, and Advanced Micro Devices have all started to adopt this viewpoint.
3. Computers that are not performing any tasks must be placed in sleep mode by the power options.
4. Computer-related tasks should be completed in contiguous, concentrated blocks of time to save money. Hardware should be turned off at other times.
5. Compared to standard CRT screens, level board screens use less power. By allowing a screen to go into standby mode without using screen savers, energy investment funds are increased.

6. Strong state drives differ from hard disc drives in that they store data in DRAM or flash memory.
7. Another excellent practice is to print only what is necessary and to use recycled content paper whenever it is practical. Today's printers support two-sided printing and cascading printing techniques.
8. It is crucial to develop PCs that can run on low power from unconventional energy sources, such as solar energy and bicycle acceleration.
9. Sharing desktop software, sharing terminals, and reusing old video cards are all examples of energy-efficient display techniques.
10. Vendors of IT – Concerns about customers leaving, new opportunities for revenue generation, and a desire to demonstrate corporate social responsibility have led IT vendors to adopt various green standards into their operational procedures.
11. Eco-friendly approach – PVC, cadmium, lead, and bromine-induced flame inhibitors, which are used to make computers, are just a few examples of the e-waste materials that electronics manufacturers are working to reduce. To achieve this, bioplastics, which are plant-based polymers and need less energy and fuel to make than conventional plastics, can be used in place of plastic that contains petroleum [22]. Keeping these bio-plastic computers cold enough to avoid melting and consequent damage is another difficulty. Landfill operators should keep an eye on them by making the most use of the equipment by tuning up and mending items as needed in order to make such a procedure quick and economical.

Table 6.1 lists some hazardous metals and chemicals that are present in computer wastes and their impacts on health, which may be slowed by utilizing environmentally friendly methods of green computing,

6.11.2 Green Computing Solutions

Following are the solutions that help with green computing in order to reduce e-hazard production and secure the environment.

1. **Create long-term green computing strategies** – All those connected to the industries, from the highest ranks to the lowest, must actively participate in order for this to happen. A list of mandatory rules, governmental regulations, green recommendations, and a list of both recyclable and non-recyclable items must be included in organizational policies and checklists [23]. In order to eliminate e-waste from organizations, best practices and strategies should focus on reducing the use of non-conventional resources, including paper used.
2. **Reusing and recycling** – Electronic equipment that is no longer needed or wanted should be disposed of in a convenient and environmentally friendly way. Environmentally harmful emissions can be produced by the toxic

TABLE 6.1
Effects of IT Devices and Wastes on Health

Metal/Compound	State	Used in Mentioned Computing Hardware	Health Implications	Route
Cadmium	Found in the solid state. Its pyrolysis produces poisonous fumes.	Resistors, nickel-cadmium batteries, screens, lasers	Damage to the kidneys, lungs, bones, muscles, chills, fever, nausea, and vomiting, as well as a decline in memory and cognitive function and neuromotor impairment.	Inhalation, ingestion
Lead	Found in the solid state. Lead oxide is formed when burnt/heated.	CRT monitors, printed boards, polyvinyl chloride formulations	Because lead is rapidly absorbed into the bloodstream through the mucous membranes of the lungs, skin, and abdomen, lead poisoning, high blood pressure, liver damage, kidney damage, poor brain development, and lifelong nerve damage in children are all possible effects.	Inhalation, ingestion
Mercury	Found in the liquid and gaseous state.	Fluorescent tubes and flat-screen monitors	This substance is harmful to the nervous, immune, and digestive systems, whether alone or in methylated forms. Other internal organs are also impacted, such as the kidneys and lungs. The eyes, skin, and digestive system are all adversely affected by mercury salts.	Inhalation, ingestion
Beryllium oxide	Found in the solid and liquid state.	Thermal grease for heat sinks; processor, power transistors, vacuum tubes	Causes irritation of skin, throat, nose, and lungs.	Inhalation contact
Barium	Found in the solid and liquid state.	CRT monitors	Poisoning	Inhalation ingestion, contact
Zinc	Found in solid state.	CRTs, metal coatings	Poisoning, vomiting, dizziness, diarrhoea, stomach pain.	Inhalation, ingestion

FIGURE 6.6 Prioritizing e-waste as important as other waste.

metals and pollutants found in computers [24]. Never throw away electronic waste in a landfill. Instead, recycle them through manufacturer services like neighbourhood recycling centres, or donate still-operational computers to a non-profit organization (Figure 6.6).

3. **Purchase environmentally friendly and ethical goods** – Make purchases of green products that are safe for the environment and you. The environmental damage brought on by energy consumption is lessened thanks to these products. Due to this, consumers ought to be encouraged to purchase eco-friendly goods. Designing green products requires the establishment of precise and definite standards. To make products that are better for the environment, manufacturers must be involved as well as given due credit. Purchase products registered by the Electronic Product Environmental Assessment Tool (EPEAT). The non-profit Green Electronics Council has been promoting EPEAT as a purchasing tool.
4. **Lowering the use of paper** – There are numerous strategies for eliminating paper consumption. Today, all tasks can be completed on the accumulator thanks to computers, which are more popular than anything else. E-mail, free messaging, and other sophisticated forms of communication have brought communication right to your door. Additionally, since more businesses are attempting to go paperless on a daily basis, additionally, the industries are also saving paper.
5. **Power-saving measures** – All electronic devices show evidence of energy consumption that comes from non-renewable energy sources. Adopt the appropriate strategies and methods to save energy so that it can be used when it is actually needed.

6.11.3 Green Computing Practices

Instead of focusing on faster analysis, quicker computations, and more complicated problem solving, green computing today aims to improve energy efficiency, minimize the consumption of electronic devices, reduce e-waste, and employ non-toxic materials in the creation of electronics. As the practice of employing green computing resources efficiently gains significant traction, the objectives of avoiding the use of hazardous chemicals, increasing energy efficiency throughout the product's lifespan, and encouraging the recyclability or biodegradability of outdated goods and manufacturing waste are being accomplished. Green computing practices are listed as follows:

1. **Virtualization** – It is the mental image of computer resources, such as when several logical computer systems are operated on a single physical hardware configuration. A system administrator can use virtualization to combine numerous physical systems into virtual computers running on a single, powerful system, which reduces the amount of power and cooling required, increases revenue, and decreases expenses over time. Reducing the number of hardware components and moving to green computing systems decreases the energy costs for cooling and running hardware, as well as reduces carbon dioxide emissions and boosts energy efficiency.
2. **Algorithm efficiency** –There are many performance advantages and disadvantages when writing programmes, and the effectiveness of algorithms affect how many resources are needed on the computer for any given computing function. By reducing the strain on the hardware, algorithms like the computation and transmission rate based algorithm (CTRB), slack reduction algorithm (SRA), integer-bit power allocation algorithm, and duEDF algorithm enable intelligent computing to function organically.
3. **Implementation of enhanced data centre cooling methods** –This is accomplished by enhancing the cooling system in the data centre and stopping significant energy leaks. Adhering to best practices for data centre layout and rack and server configuration can produce effective data centres [25]. Raised flooring, relocating cooling units closer to servers to concentrate just on cold air where it is required, switching between hot and cool server corridors, and utilizing water-based air conditioning systems are all effective techniques.
4. **Power management** – The ACPI, an open industry standard, allows an operating system to directly handle the hardware's power-saving capabilities (Figure 6.7 and Figure 6.8). This lets a system to automatically switch off equipment like displays and hard drives after a set amount of idleness. A system can also enter a state known as hibernation in which most of its components, including the CPU and system RAM, are shut down. Advanced power management, a previous Intel–Microsoft standard that gave the BIOS of the computer control over power management operations, was replaced by ACPI.
5. **Power supply** – Power supply units (PSUs) for desktop computers typically have an efficiency of 70%–75%, the heat produced by the leftover energy. An industry programme known as 80 PLUS certifies PSUs that operate at least 80% effectively. Often, these models are drop-in replacements for older, less cost-effective PSUs with the same form factor [26]. As of July 20, 2007, all new desktop PSUs with Energy Star 4.0 certification must be at least 80% efficient.

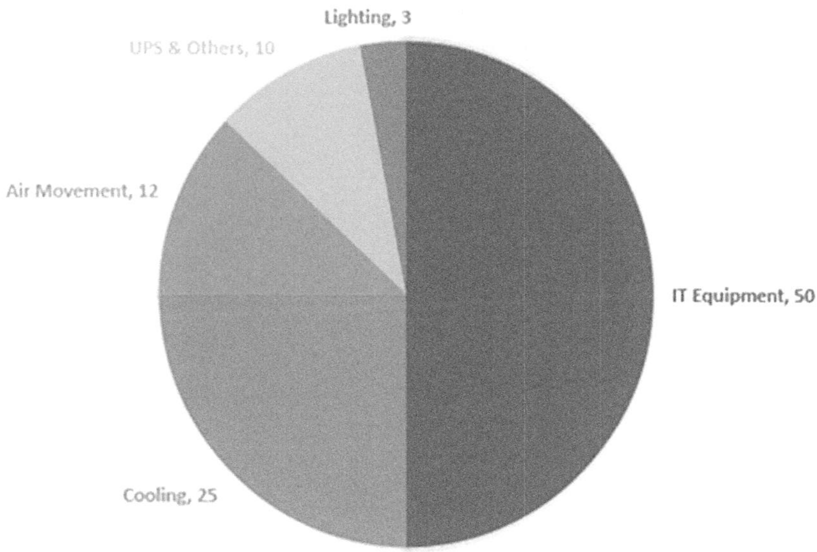

FIGURE 6.7 Computer power wattage in watts (W).

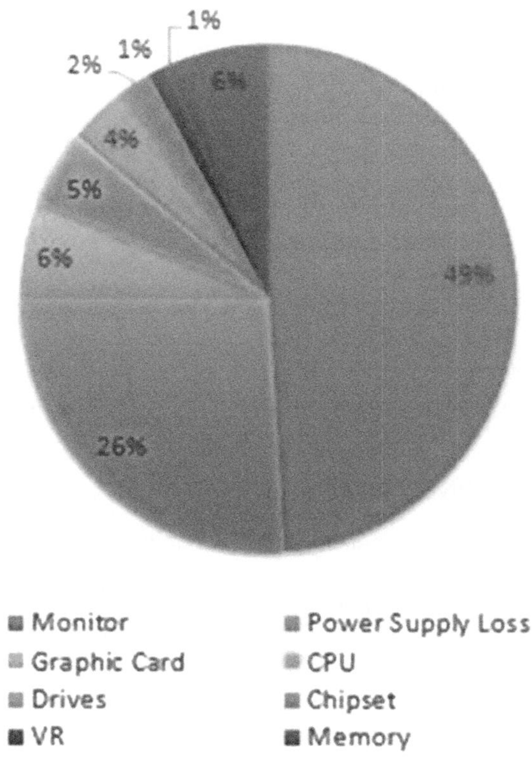

FIGURE 6.8 Computer component wattage.

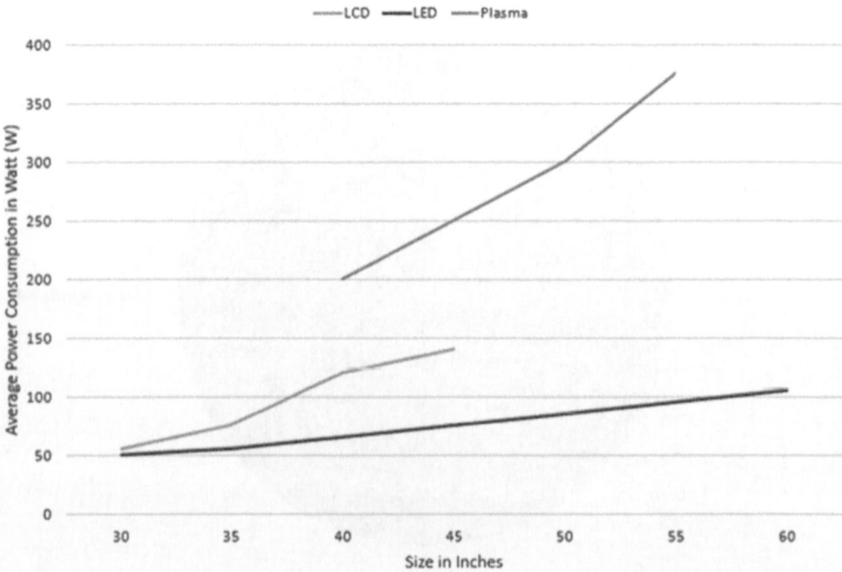

FIGURE 6.9 Power consumption as per display.

6. **Storage** – Hard disc drives with smaller form factors frequently have lower power consumption per gigabyte than drives with larger physical dimensions. Data is stored in flash memory or DRAM on solid-state drives, not on hard disc drives. Power consumption for limited power flash-based devices may be slightly decreased due to the absence of moving parts. Even at small sizes, DRAM-based SSDs might utilize more power than conventional hard drives (e.g., 4GB IRAM uses more power and space than laptop drives) [27].
7. **Display** – Cold-cathode fluorescent lighting is frequently used in LCD monitors to illuminate the display. In some more modern displays, a grid of light-emitting diodes (LEDs/OLEDs) takes the role of the fluorescent bulb, which lowers the electricity consumption of the display (Figure 6.9).

6.12 IMPLEMENTING STRATEGIES FOR GREEN COMPUTING

The following are the various strategies involved in implementing green computing,

1. **Create a strategy for long-term green computing** – Business executives should be consulted on the elements that should be contained in such a strategy, such as organizational policies and checklists. Such a plan ought to contain guidelines for recycling, suggestions for getting rid of outdated technology, regulatory requirements, and recommendations for purchasing environmentally friendly computer gear [28]. Best practices and policies for

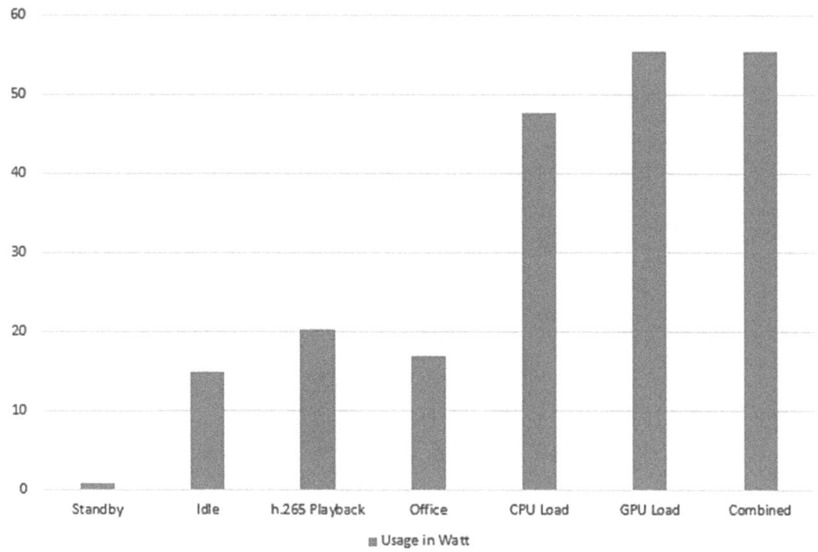

FIGURE 6.10 Computer power consumption in watts (W).

green computing should address power usage, paper consumption reduction, suggestions for new equipment, and recycling of used equipment. Implementation and communication should be covered in organizational policies.

2. **Calculating the present carbon footprints of IT components** – It is crucial to understand the company's position with regard to the carbon footprint created by information technology services (Figure 6.10 and Figure 6.11). Establish a carbon footprint benchmark as soon as possible [29]. Compare the power usage in the IT centre to the industry's current power efficiency metrics and standards.
3. **Electronics should be disposed of in a useful and eco-friendly way** – Computers contain metals and other contaminants that can emit toxic gases into the atmosphere. Never throw away electronic waste in a landfill. Instead, recycle them using local recycling facilities or manufacturer programmes, such as HP's Planet Partners recycling service. Alternatively, give operating computers to a charitable organization.
4. **Utilization of more powerful computer applications** – Your electronic device systems can better handle inefficiencies by utilizing more potent computer applications. Additionally, faster software prevents the servers from constantly running at full capacity, using less power [30]. The amount of energy used and carbon emissions could be reduced in a similar manner if computer applications were only run faster.
5. **Save energy** – When you know you won't be using your computer for a while, turn it off. When inactive for shorter amounts of time, activate power

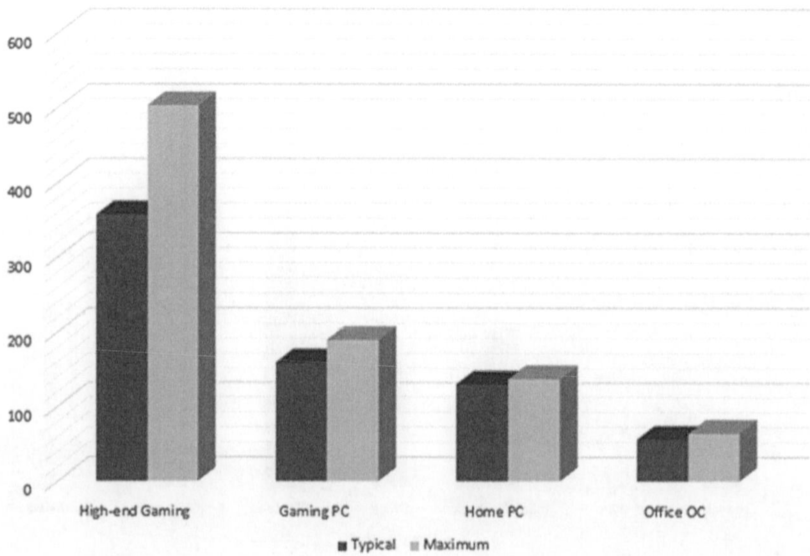

FIGURE 6.11 Tablet power consumption in watts (W).

management features. Thanks to power management, computers and monitors can enter low-power states when not in use [31]. Simply touching a key on the keyboard or moving the mouse causes the computer or monitors to emerge from their low-power sleep mode in a matter of seconds.

6. **Enhancing business performance with sustainable IT practices** – Ensure the push for green IT integrates with your company's overall operations. Even better, make sure that the objectives of the business and environmentally friendly IT are complementary. You can accomplish both green policies and financial objectives by doing this.

The aforementioned strategies can be easily implemented by organizations, individual users, and the government to reduce electronic hazards and safeguard the environment.

6.12.1 Industry and Government Schemes in Favour of Green Computing

In order to ensure adherence to green computing, the industry and governments have established the following measures.

1. **Government initiatives** – Numerous governmental organizations have kept up the implementation of rules and standards that promote sustainable IT. In October 2006, the Energy Star programme underwent revisions to include more stringent efficiency standards for computer hardware [32]. To comply with directives 2002/95/EC (RoHS) on the reduction of hazardous

substances and 2002/96/EC (WEEE) on electrical and electronic waste, all electronic devices put on the market after July 1, 2006, had to be devoid of flame retardants including PBBs and PBDEs, as well as heavy metals. The guidelines mandated that manufacturers be in charge of collecting and recycling used equipment (the producer responsibility model).

2. **Industry initiatives**
   a. **Climate Savers Computing Initiative (CSCI)** – It is an initiative to lower the amount of electricity used by PCs while they are active and idle. The CSCI offers information on lowering PC power consumption as well as a catalogue of eco-friendly goods from its member companies. It first began on June 12, 2007.
   b. **Green Computing Impact Organization Inc. (GCIO)** – A non-profit organization called GCIO was founded with the goal of helping computer users make environmentally responsible decisions. This aim is accomplished through educational programmes, community projects, and free auditing services. The GCIO Cooperative, a network of IT executives who care about the environment, serves as the foundation of the organization. Members of this cooperative dedicate their time, money, and purchasing energy to spread awareness of green computing, increase its uptake, and boost its effectiveness.
   c. **Green Electronics Council** – In order to make green computing system purchases easier, the Green Electronics Council provides the EPEAT. The council assesses computer hardware using 28 criteria that gauge a product's sustainability and efficiency. Executive Order 13423, signed by President George W. Bush on January 24, 2007, calls for all federal agencies in the United States to use EPEAT when making computer system purchases.
   d. **The Green Grid** – It is a global consortium working to improve data centres and business computing ecosystems' energy efficiency. In February 2007, it was founded by a number of significant business players in the sector, including AMD, APC, Dell, HP, IBM, Intel, Microsoft, Rackable Systems, SprayCool, Sun Microsystems, and Vmware. Since then, the Green Grid has increased to hundreds of participants, along with end users and governmental bodies, all of whom are committed to enhancing data centre effectiveness [33].

3. **Niti Aayog Initiative** – It is important to assess the situation and plot the future course as India's green energy industry gets ready to take off. Reforms in this industry have been ongoing for more than ten years, and in the last five years, installed capabilities from renewable electricity have doubled, recently reaching 100 GW. India still depends on imported oil and gas, despite the fact that this is unquestionably a notable improvement. In accordance with the United Nations' Sustainable Development Goal 7 "Ensure the availability of affordable, reliable, sustainable, and modern energy for all"," the prime minister made a commitment to make India

> energy independent by 2047. In order to achieve this, policy implementation must be challenged, and it must begin at the local level rather than in highly developed areas [8].

The prime minister of India acknowledged the importance of "Sabka Prayas", or participation from businesses, local authorities, and communities. The potential for change in local communities is enormous. The Aspirational Districts Programme, which was established under the auspices of "Sabka Saath, Sabka Vikas", aims to involve everyone in the country's economy. Without appropriate policy interventions which can reach the grassroots, it is very challenging to implement an all-sharing growth model in a nation as economically diverse as India. These interventions support the development of robust systems that give priority to subnational contexts and needs, and use the distinctive delta ranking system to track long-term progress toward sustainable development.

The United Nations Development Programme (UNDP) emphasizes the application of green technology for the construction of fundamental infrastructure in Aspirational Districts. It features Goalpara, Assam, which built 183 km of the green road using geogrid technology and recycled plastic. In fact, it is India's first district to build a green road. The establishment of more than 400 dwellings in the district with connectivity to all-weather roads has resulted in a significant improvement of the delta. The poorest get the greatest advantage and can make the most delta change as the country advances on its path to a greener tomorrow.

Electricity is a part of the basic infrastructure under the Aspirational District Programme. The interventions should start with solar energy because it is a mature and feasible green energy technology. The geography of India and the cutting-edge solar cells work in favour of solar energy. Even small solar farms can be constructed. This improves accessibility and lowers the price of transmission. Despite these benefits, solar energy adoption has been gradual [34]. The programme will fuel India's ambitions to become a global leader in green energy and carry out the United Nations' 2030 Agenda thanks to its expanded scope in energy transitions. It not only acts as a lighthouse for our green energy revolution but also offers a replicable model for the rest of the world.

Table 6.2 lists policies implemented in three countries related to GHG emissions, energy consumption, data centres, and e-waste sector policies.

6.12.2 Sustainable Development with Green Computing in Post-COVID Era

Green computing and environmentally friendly IT are related to sustainable development. The post-COVID era is characterized by uncertainty, particularly in the ecosystem of public health and safety. This necessitates a review and strengthening of the current sustainable development–related laws, regulations, policies, guidelines, and controls. Accordingly, environmental deterioration and climate change have a higher propensity to result in more disturbing ecological catastrophes in the

TABLE 6.2
Green Computing Policy Implementation in Three Leading Countries

	Country		
Category	**United States of America**	**United Kingdom**	**India**
Greenhouse gas (GHG) emissions	The US Environmental Protection Agency (EPA) tracks GHG emissions.	The United Kingdom government established the Climate Change Act to control GHG emissions.	To lessen the effects of climate change, the National Action Plan on Climate Change was introduced.
Energy consumption	Utilizing renewable energy to meet energy needs and lessen the impact of GHG.	Energy Companies Obligation (ECO) and the Green Deal are two energy efficiency programs.	To improve energy efficiency, the National Mission for Enhanced Energy Efficiency was created.
Data centres	Energy services performance contracts (ESPCs) and generation and utility energy service contracts (UESCs) for data centre efficiency improvement.	Consolidating data centres is a policy that has been implemented to lower the energy used to cool facilities.	The Department of Electronics and Information Technology directs sustainable data centre practices.
E-waste	E-waste is governed by the National Computer Recycling Act and the Resource Conservation and Recovery Act.	E-waste is governed by the 2014 Waste Electrical and Electronic Equipment (WEEE) regulations.	E-waste is governed by the E-waste (Management & Handling) Rules, 2011.

post-COVID era [35]. The key objectives of sustainable development are supposed to be social inclusion, environmental sustainability, and economic development. With the advent of technologies such as Wireless Networks, Delay Tolerant Networks, Cloud Computing, Internet of Things [3736–40]. Green IT is a significant driver of sustainable development due to its focus on a healthier, safer, and more sustainable environment that encourages the general well-being of all individuals within the natural ecosystem. The relationship between green computing and sustainable development is summarized in Table 6.3.

6.13 FUTURE OF GREEN COMPUTING

As the public conversation on carbon emissions, global warming, and climate change continues, energy issues will be much more important in the coming days. Computers, gadgets, and electronic devices rule the 21st century. We need to reconsider our assumptions if we believe that electronic computer devices are

TABLE 6.3
Contribution of Green IT to Sustainable Development

Sustainable Development Goal	Contribution of Green Computing to Sustainable Development
No poverty	The average worker's productivity, health, and safety would all be guaranteed by green computing's commitment to a cleaner workplace
Zero hunger	Green computing would improve crop yields and food security by reducing soil contamination and pollution.
Good health and well-being	Human health and well-being are at risk from the greenhouse gases released by computers and e-waste. The social well-being of the populace would improve if these threats were eliminated.
Quality education	A clean and safe environment has a positive impact on teaching and research. When conditions are like this, education creates knowledgeable professionals who could aid in national and international development.
Gender equality	Water, air, and soil contamination would be removed by green computing, and health improved. Women and girls would no longer be required to perform household chores like finding water as a result of a cleaner environment and access to water.
Clean water and sanitation	Leaches that contaminate groundwater could be avoided with proper e- waste disposal and management of carbon load from large computing environments.
Affordable and clean energy	For computers and related devices, green computing favours clean, renewable energy.
Decent work and economic growth	Workplaces that are safer and cleaner motivate employees, which increases productivity and, consequently, economic growth.
Industry, innovation, and infrastructure	Green computing is a cutting-edge philosophy that encourages harmony and a pollution-free industrial environment. It promotes the creation of environmentally friendly tools and infrastructure.
Reducing inequality	No group or community should be forced to bear an excessive share of the negative effects of pollution, thanks to effective pollution governance and actions.
Sustainable cities and communities	In local and urban areas, a healthy, toxic-free environment encourages sustainable production and economic activity.
Responsible consumption and production	Green computing encourages firms to adopt greener production and manufacturing processes, which would decrease consumers' constant demand for newer devices due to the falling utility value of existing devices, driving sustainable consumption and manufacturing activities.
Climate action	Green computing emphasizes the use of renewable energy sources with cleaner energy footprints and the safe disposal of computer-based wastes that could harm the environment and increase the cost of climate change.
Life in water	Electronic waste dumps contaminate underground waters, killing important aquatic organisms as a result. Green computing promotes ethical e-waste disposal.

(*Continued*)

TABLE 6.3 (CONTINUED)
Contribution of Green IT to Sustainable Development

Sustainable Development Goal	Contribution of Green Computing to Sustainable Development
Life on land	The constant maintenance of a safe, healthy, and clean environment is one of green computing's main objectives.
Peace, justice, and strong institutions	In all socioeconomic spheres, impartially enforcing green computing policies and regulations would foster peace and safety, fostering confidence and trust in public institutions.
Partnerships for the goals	National, regional, and international levels should all be affected by green computing ideals and policies. Such partnerships would promote the unity of purpose and have the greatest impact on the regulation of electronic waste disposal as well as the adoption of more environmentally friendly production techniques. The outcome would have an impact on all ecosystem spheres.

non-polluting and don't use much energy. Tons of carbon emissions will be saved annually for every unit of energy saved in computing and computer hardware [41, 42]. Given the widespread use of information technology, the business demands initiating a revolution by adopting green practices in a way that no other sector has. In server farms where the HVAC and computers are housed together, inefficient CPUs are a double whammy since they use too much electricity and increase the necessity for cooling. Waste heat also contributes to reliability issues because high temperatures greatly increase the frequency of CPU crashes. Years have been spent working to rid computers of this inefficiency. Power supplies are also notoriously bad, typically only being 7% efficient [43–45]. Furthermore, since the power supply is the source of power for the entire computer, efficiency is dependent on the power supply. New power supply innovations that operate at 80% efficiency or higher are helping to fix this.

6.14 CONCLUSION

The ineffective use of technology has made it a significant contributor to global warming. Before it's too late, we must feel a strong urge to protect our environment. Green computing is the evolving strategies of efficient utilization of IT and computing assets while improving and managing overall operations. Due to the increasing concern about the environment, green computing has received a lot of attention. Green computing strategies cut energy use and carbon dioxide emissions. Humans can avoid paying for irrational energy and resource use by implementing green computing strategies. Governmental regulations are constantly shifting to promote recycling, and green computing addresses concerns such as chemicals in computers that are deemed dangerous. To manufacture, use, and dispose of items, energy-efficient resources are used. Green computing is playing an increasingly important role as computers are often used ever more extensively. Utilizing green computing has different strengths and weaknesses, but the advantages surpass the drawbacks when evaluated against the limitations, thus it is essential.

REFERENCES

1. Saxena, Shailesh, Mohammad Zubair Khan, and Ravendra Singh. "Green computing: An era of energy saving computing of cloud resources." *International Journal of Mathematical Sciences and Computing (IJMSC)* 7.2 (2021): pp 42–48.
2. Mittal, Pardeep, and Navdeep Kaur. "Green computing–need and implementation." *International Journal of Advanced Research in Computer Engineering & Technology (IJARCET)* 2.3 (2013): 1–4.
3. Chowdhury, Srinjoy Nag, Kshitij Manohar Kuhikar, and Akshay Agnihotri. "Green computing: An overview with reference to India." *International Journal of Electrical, Electronics and Computer Systems* (2015).
4. Dhaini, Mahdi, et al. "Green computing approaches-A survey." *Informatica* 45.1 (2021).
5. Ali, Md Akkas, B. Balamurugan, and Vandana Sharma. "IoT and blockchain based intelligence security system for human detection using an improved ACO and heap algorithm." In *2022 2nd International Conference on Advance Computing and Innovative Technologies in Engineering (ICACITE)*. IEEE, 2022.
6. Durgalakshmi, K., et al. "Utilization of reduced switch components with different topologies in multi-level inverter for renewable energy applications-A detailed review." In *2022 5th International Conference on Contemporary Computing and Informatics (IC3I)*. IEEE, 2022.
7. Podder, Suplab Kanti, and Debabrata Samanta. "Green computing practice in ICT-based methods: Innovation in web-based learning and teaching technologies." *International Journal of Web-Based Learning and Teaching Technologies (IJWLTT)* 17.4 (2022): 1–18.
8. Bansal, K., Kurle, P., and Chawla, P. Green IT towards Digitalizing India. *International Journal on Recent and Innovation Trends in Computing and Communication*, 6(1), 12–20.
9. Jindal, Gaurav, and Manisha Gupta. "Green computing "future of computers"." *International Journal of Emerging Research in Management & Technology* (2012): 14–18.
10. Ahmad, Shazib, Saksham Mishra, and Vandana Sharma. "Green computing for sustainable future technologies and its applications." In *Contemporary Studies of Risks in Emerging Technology, Part A*. Emerald Publishing Limited, 2023. 241–256.
11. Mukta, Tamanna Akther, and Iqbal Ahmed. "E-waste management strategies for implementing green computing." (2021).
12. Modanval, R. K., Rakesh Kumar, S., Gayathri, N., Ramesh, C., Kumar, A., & Sharma, S. K. (2021). Impact of green computing in shaping education. *Green Computing in Smart Cities: Simulation and Techniques*, 171–188.
13. Debnath, Biswajit, Reshma Roychoudhuri, and Sadhan K. Ghosh. "E-waste management–a potential route to green computing." *Procedia Environmental Sciences* 35 (2016): 669–675.
14. Sharma, V., Balusamy, B., Thomas, J. J., and Atlas, L. G. (Eds.). (2023). *Data Fabric Architectures: Web-Driven Applications*. Walter de Gruyter GmbH & Co KG.
15. Shibly, A. "Green computing:-Emerging issue in IT." (2015).

16. Saha, Biswajit. "Green computing." *International Journal of Computer Trends and Technology (IJCTT)* 14.2 (2014): 46–50.
17. Monga, Hitesh. "Green computing: Approaches and its implementations."
18. Okewu, Emmanuel, et al. "Optimizing green computing awareness for environmental sustainability and economic security as a stochastic optimization problem." *Sustainability* 9.10 (2017): 1857.
19. Khasawneh, Ahmad M., et al. "Green computing in underwater wireless sensor networks pressure centric energy modeling." *IEEE Systems Journal* 14.4 (2020): 4735–4745.
20. Kumar, Sushil, Omprakash Kaiwartya, and Abdul Hanan Abdullah. "Green computing for wireless sensor networks: Optimization and Huffman coding approach." *Peer-to-Peer Networking and Applications* 10 (2017): 592–609.
21. Nwankwo, Wilson, S. A. Olayinka, and Kingsley E. Ukhurebor. "Green computing policies and regulations: A necessity." *International Journal of Scientific & Technology Research* 9.1 (2020): 4378–4383.
22. Sarkar, Subhadeep, and Sudip Misra. "Theoretical modelling of fog computing: A green computing paradigm to support IoT applications." *Iet Networks* 5.2 (2016): 23–29.
23. Gai, Keke, et al. "Dynamic energy-aware cloudlet-based mobile cloud computing model for green computing." *Journal of Network and Computer Applications* 59 (2016): 46–54.
24. Debnath, Biswajit, Reshma Roychoudhuri, and Sadhan K. Ghosh. "E-waste management–a potential route to green computing." *Procedia Environmental Sciences* 35 (2016): 669–675.
25. Ojo, Adedapo Oluwaseyi, Murali Raman, and Alan G. Downe. "Toward green computing practices: A Malaysian study of green belief and attitude among Information Technology professionals." *Journal of cleaner production* 224 (2019): 246–255.
26. Farhan, Laith, et al. "Towards green computing for Internet of Things: Energy oriented path and message scheduling approach." *Sustainable Cities and Society* 38 (2018): 195–204.
27. Kumar Kashyap, Pankaj, et al. "Green computing in sensors-enabled internet of things: Neuro fuzzy logic-based load balancing." *Electronics* 8.4 (2019): 384.
28. Han, Guangjie, et al. "Resource-utilization-aware energy efficient server consolidation algorithm for green computing in IIOT." *Journal of Network and Computer Applications* 103 (2018): 205–214.
29. Sengupta, Gaurav. "Green computing–new perspective of efficient usage of energy and reduction of E-Waste." *Journal of Technology and Innovation in Tertiary Education* 2.1 (2019): 11–16.
30. Pushpalatha, N., et al. "A succinct summary of the solar MPPT utilizing a diverse optimizing compiler." In *2022 5th International Conference on Contemporary Computing and Informatics (IC3I)*. IEEE, 2022.
31. Singh, Shivam. "Green computing strategies & challenges." In *2015 International Conference on Green Computing and Internet of Things (ICGCIoT)*. IEEE, 2015.
32. Juyal, V., A. V. Singh, and R. Saggar. "Message multicasting in near-real time routing for delay/disruption tolerant network." In *2015 IEEE International Conference on Computational Intelligence & Communication Technology*. IEEE, 2015, February. 385–390.
33. Juyal, V., R. Saggar, and N. Pandey. "Clique-based socially-aware intelligent message forwarding scheme for delay tolerant network." *International Journal of Communication Networks and Distributed Systems* 21.4 (2018): 547–559.
34. Sharma, Vandana, et al., eds. *Data Fabric Architectures: Web-Driven Applications*. Walter de Gruyter GmbH & Co KG, 2023.

35. Ahmad, Shazib, Saksham Mishra, and Vandana Sharma. "Green computing for sustainable future technologies and its applications." In *Contemporary Studies of Risks in Emerging Technology, Part A*. Emerald Publishing Limited, 2023. 241–256.
36. Juyal, V., R. Saggar, and N. Pandey. "On exploiting dynamic trusted routing scheme in delay tolerant networks." *Wireless Personal Communications* 112 (2020): 1705–1718.
37. Singh, Anamika, et al. "Blockchain: Tool for controlling ransomware through pre-encryption and post-encryption behavior." In *2022 Fifth International Conference on Computational Intelligence and Communication Technologies (CCICT)*. IEEE, 2022.
38. Juyal, V., N. Pandey, and R. Saggar. "Opportunistic message forwarding in self organized cluster based DTN." In *2017 International Conference on Infocom Technologies and Unmanned Systems (Trends and Future Directions) (ICTUS)*. IEEE, 2017, December. 497–502.
39. Poongodi, T., Ramya, S. R., Suresh, P., and Balusamy, B. (2020) Application of IoT in green computing. *Advances in Greener Energy Technologies*, 295–323.
40. Juyal, Vandana, Nitin Pandey, and Ravish Saggar. "A heuristic light weight security algorithm for resource constrained DTN routing." In *2016 IEEE International Conference on Computational Intelligence and Computing Research (ICCIC)*. IEEE, 2016.
41. Juyal, V., R. Saggar, and N. Pandey. "An optimized trusted-cluster–based routing in disruption-tolerant network using experiential learning model." *International Journal of Communication Systems* 33.1 (2020): e4196.
42. Balusamy, Balamurugan, Naveen Chilamkurti, and Seifedine Kadry, eds. *Green Computing in Smart Cities: Simulation and Techniques*. Springer Nature, 2020.
43. Juyal, Vandana, Nitin Pandey, and Ravish Saggar. "Performance comparison of DTN multicasting routing algorithms-opportunities and challenges." In *2017 International Conference on Intelligent Sustainable Systems (ICISS)*. IEEE, 2017.
44. Maurya, Aakash, and Vandana Sharma. "Facial emotion recognition using Keras and CNN." In *2022 2nd International Conference on Advance Computing and Innovative Technologies in Engineering (ICACITE)*. IEEE, 2022.
45. Balusamy, Balamurugan, et al. "Bio-inspired algorithms for cloud computing: A review." *International Journal of Innovative Computing and Applications* 6.3–4 (2015): 181–202.

7 Implementation Strategies for Green Computing

Kuldeep Singh Kaswan, Jagjit Singh Dhatterwal, Santar Pal Singh, Vandana Sharma, and Balamurugan Balusamy

CONTENTS

7.1 Introduction 136
7.2 Green Building 136
7.3 Innovative Green Building: The Impact of Renewable Energy Sources and Techniques 137
 7.3.1 Solar Power 137
 7.3.2 Biodegradable Materials 138
 7.3.3 Green Insulation 138
 7.3.4 Smart Appliances 138
 7.3.5 Cool Roofs 138
 7.3.6 Sustainable Resource Sourcing 139
 7.3.7 Designing for Low or No Energy Consumption 139
 7.3.8 Electrochromic Smart Glass 139
 7.3.9 Systems for Efficient Use of Water 140
 7.3.10 Sustainable Indoor Environment Technologies 140
 7.3.11 Power-Sufficient Structures 140
 7.3.12 Rammed Earth Brick 140
7.4 Eco-Friendly Building Power Savings 141
7.5 Buildings That Combine Energy-Saving and Eco-Friendly Technologies to Raise the Standard for Green Design 142
 7.5.1 Systems That Combine Photovoltaic and Thermal (PV/T) Energy 144
 7.5.2 Construction of Atriums Using Fresnel Lenses 144
 7.5.3 Integration of Wind/Solar Systems 144
 7.5.4 The Financial Impact of Renewable Energy Technologies 145
7.6 Renewable Energy Technologies and the Long-Term Viability of Eco-Friendly Buildings 146
7.7 Conclusion 146
References 147

DOI: 10.1201/9781003348313-7

7.1 INTRODUCTION

One definition of "green building" is "a structure that is created, constructed, built, managed, reused, or refurbished in a resource-efficient and ecological manner". In environmental terms, a sustainable system is one that does not deplete its resources at a quicker rate than they can be naturally supplied and hence continues to function. According to standard economic theory, a fiscally sound economic policy is one in which outgoings are less than or equal to receipts (Kounatze, 2009). To put it in terms of a shared language, a system of governance that is able to endure is one in which its members are collectively empowered, leading to the creation of a more effective whole. One's personal choices have consequences for the whole planet, as is becoming more obvious. Green building has earned worldwide attention for its emphasis on energy efficiency and conservation (Dhatterwal, 2022). There are four main considerations when it comes to renewable energy production in a green building: first and foremost, the lower level plan before actual prototype and renovation; the use of low-energy construction methods during building works; the use of electricity types of equipment for the low operational energy necessity; and the superior human mix of renewable electricity advancements for various applications (Dedrick, 2010).

We have considered the ecological and monetary effects these proposals might have on the next centuries. These days, it's conceivable for buildings all over the world to be as functional and comfortable as possible since "home" and "habitat" have merged into one term. The building sector is expanding at an accelerated pace, consuming 30%–40% of the world's key resources. Currently, the construction sector is the third largest user of fossil fuels in terms of energy consumption, behind the industrial and agricultural sectors. The Asia-Link initiative was established by the European Commission to disseminate and enhance expertise in the zero-energy approach to a sustainable built environment. In this sustainable architecture scheme, renewable energy sources are incorporated into the structure itself to be used for multiple functions, including heating, cooling, and powering the building's appliances (Wabwoba, 2013). Across the board, the building's operating electricity consumption is negligible. Thus, evaluating the sustainability framework is essential, particularly in the construction industry, for global sustainable development (Ahmad et al., 2023). The primary goals and objectives of the goals of ecological sustainability are to lessen demands on finite resources like water, natural resources, and electricity; to forestall environmental destruction from transportation systems and advantages throughout their entire existence cycles; and to create settings that make optimal use of renewable resources like solar and water power without compromising human health and safety (Molla, 2009).

7.2 GREEN BUILDING

A natural sustainable infrastructure also has to make use of data and analysis gained from the surrounding environment and the building's adaptation in order to qualify as one (Philipson, 2010). The location must be chosen using the original data. Colours used in construction should be natural hues, not those manufactured in a lab. In natural construction practices, it is necessary to use the resources that nature already provides. It is strongly suggested to utilize natural materials instead of synthetic items since creating synthetic products requires a lot of energy. Natural materials include components that may be processed using either contemporary or historical artisan practices. The term "passive building", which may be achieved with a wide range of building materials, is a standard for energy efficiency in a building. It's possible to accept this as an example of green building design since the designers state that the inside environment doesn't need any additional heating or cooling systems, regardless of the season. This is one of the prerequisites for achieving modest energy use (Hart, 1997). For a structure to be considered passive, one criteria example is high-quality, triple-paned windows oriented toward the sun to gather energy passively. Designing for lower energy use as well as incorporating sustainable farming practices and enough natural light into indoor spaces have been shown to boost productivity in the workplace without increasing costs for builders. Changing the orientation of a building's window frames, for instance, may help take advantage of the sun's stray rays. The green building concept has an advantage of implementing solar powered equipement that are sustainbale and self sufficient as stated in (Durgalakshmi et al., 2022; Pushpalatha et al., 2022)

7.3 INNOVATIVE GREEN BUILDING: THE IMPACT OF RENEWABLE ENERGY SOURCES AND TECHNIQUES

Because they produce no harmful by-products, renewable energy sources leave less of an ecological footprint than their fossil fuel counterparts. There is no emission of greenhouse gases, thus, it's good for the planet. Renewable sources of energy may be used forever. In contrast, it is inevitable that the world's supply of nonrenewable energy will run dry (Dhatterwal, 2020). New employment opportunities will become available when the economy improves. Energy efficiency is a major factor in the design of green buildings. It's a big reason why it's been so successful. Several solutions that improve electrical power management, lower energy consumption, and increase the building's energy efficiency must be installed in a green structure. Renewable energy has the distinct benefit of having an infinite supply at its disposal (Boughton, 2009). Using renewable energy is an effective means of securing sufficient energy for everyone while simultaneously reducing our impact on the planet. Renewable energy may be derived from various natural processes, such as water, wind, biomass, the sun, and even geothermal heat.

7.3.1 Solar Power

Solar energy's potential as an eco-friendly building material has been rapidly realized. Sustainable building projects may benefit from using active solar electricity. Active solar power generates electricity and heat using solar energy functioning systems that collect solar radiation. Typical household gas and electricity use is decreased. While the upfront costs are high, the long-term benefits of fewer emissions of greenhouse gases and lower energy bills from using renewable sources more than make up for them. Another kind of solar power, "passive solar power", involves strategically placing windows and other reflective surfaces inside a building so that the sun's rays warm the interior. When the windows are open, heat and energy from the outside may enter the home, reducing the need to utilize heating systems during the colder months (Molla, 2010).

7.3.2 Biodegradable Materials

Using eco-friendly methods, such as biodegradable materials, may help ensure a building's long-term viability. Traditional building practices often lead to the accumulation of harmful chemicals and waste products, which only decompose after hundreds of years or even thousands of years. And as a by-product of disintegration, it causes pollution and environmental damage. Biodegradable materials, such as organic paints, may reduce environmental damage since they break down quickly without producing toxic by-products. Natural materials are used in sustainable construction techniques, such as exterior walls, insulation, and concrete (Mariani, 2012).

7.3.3 Green Insulation

Insulation is a major factor in the building industry. Insulates are just wall filters; thus, they don't have to be manufactured from anything fancy. The elimination of the need for high-end items made from nonrenewable resources proves that green insulation is a sustainable building and agriculture solution (Ali et al., 2022). Recycling old newspapers or denim may be a great way to save money and the environment by acting as insulation (Dhatterwal, 2021).

7.3.4 Smart Appliances

Sustainable and green techniques and using smart appliances produced from them are necessities since commercial buildings and residences account for most electricity consumption. Sustainable building practices centre on the use of energy- and water-saving equipment. Green technology includes home appliances, such as fridges, dishwashers, washing machines, and the smart grid. This technique is geared at constructing commercial and residential structures that need no additional energy to run (Davis, 1989).

7.3.5 Cool Roofs

Cool roofs, which are intended to deflect sunshine and heat, are a kind of sustainable green roofing technology. Adjusting the building's thermal emittance and heat absorption, it helps keep houses and buildings at the correct room temperature. The design is constructed using individual tiles and reflective coatings that soak up some heat while reflecting most of the sun's rays. Cool roofs may lower temperatures by more than 50 degrees Celsius on a hot summer day. Because of less need for auxiliary systems like air-conditioning, power plants release fewer greenhouse gases as a whole (Venkatesh and Davis, 2000).

7.3.6 Sustainable Resource Sourcing

Sustainable resource procurement is a prime example of environmentally responsible building practices since it prioritizes the adoption and use of recyclable and eco-friendly building materials. Agricultural by-products and wastes are often employed in the production of building materials. Collectively, the materials may be recycled or reused, or even replicated, and they are sourced ethically (Ogunyemi and Johnston, 2012).

7.3.7 Designing for Low or No Energy Consumption

Sustainable building methods often include mechanisms to reduce energy use. Wood is used in sustainable building techniques because it requires less energy to produce than other building materials like concrete or steel. Green building incorporates high-performance insulation and windows into well-ventilated, airtight structures that allow for natural airflow. Using these methods, you may decrease your home's reliance on HVAC systems. To cut down on the need for artificial lighting throughout the day, the windows are carefully positioned. Solar energy, for example, may be used to heat water and power lights, making it an essential part of low-energy and zero-energy architectural design. While the upfront investment required to construct a zero-energy building may be high, the savings accrue over time. During the colder months, the air is warmed by using the water's preheated capability. For this reason, the heat exchanger is often used in tandem with a water–air heat exchanger (WAHE). The air–air heat exchanger (AAHE) then uses this preheated fresh ventilation air and the preheated exhaust room air to produce even more heat for the area. In order to use the already heated air from the air-conditioning system, the solar water heater incorporates a heating coil into its design. Fresh air is properly delivered to different rooms of the house at different times of the day and night, according to the individual's preferences, through a subterranean WAHE pump and AAHE equipped with smart control sensors. Such dwellings in Belgium are airtight because they are coupled with sophisticated air-handling devices, which heat the air in the rooms to the right temperature (Apulu and Latham, 2009).

7.3.8 Electrochromic Smart Glass

Technologies like electrochromic smart glass are also part of sustainable building practices. This is cutting-edge tech. Electrochromic smart glass is utilized in the summer to block the sun's intense heat. Smart glass utilizes microscopic electric impulses to energize the window to a small extent, altering the strength with which it reflects solar energy. Because smart glass is integrated into the buildings' management systems, users may choose the level of solar radiation blockage they like. Saving money on HVAC systems is a major advantage to commercial and residential buildings. Smart glass is now in the research phase but will soon be widely employed as clever energy-saving technology in environmentally friendly structures (Alena and Libor, 2012).

7.3.9 Systems for Efficient Use of Water

The sustainable construction toolbox includes several water-efficient technologies. These methods ensure that wastewater is recycled and kept in good condition for non-potable applications including indoor plumbing and vehicle washing. Dual plumbing, for instance, boosts the feasibility of on-site water recycling while reducing the number of people using the sewers. On the other side, water conservation allows for the collection and storage of water for use in various ways down the road. Technology for reducing water waste and increasing water efficiency is an important part of green building. The freshwater crisis is helped by the 15% decrease in urban water waste (Page and Hill, 2008).

7.3.10 Sustainable Indoor Environment Technologies

The well-being and security of the future occupants of a building must be given priority throughout its construction. Sustainable interior innovations are a requirement of green building. Materials that are nontoxic, hazardous-free, and resistant to moisture are utilized to guarantee compliance with green safety requirements. Bamboo, wood, and cork are all examples of eco-friendly materials that do not include any potentially harmful chemicals. Improved indoor air quality is achieved using low-volatile organic compound materials, which also reduces exposure to carcinogenic chemicals, including vinyl, lead, and phenol-formaldehyde.

7.3.11 Power-Sufficient Structures

Sustainability in building practices is the future, and it may be used to create structures that need no external power source. Related to the actualization of the self-powered, zero-energy building project, self-sufficient structures are designed to generate enough electricity to meet their own energy demands and export any surplus to the grid (Raj et al., 2022). Many tall buildings have wind turbines installed on their roofs so that they may generate their own electricity. The required power for the structure is generated by the turbine blade, which is powered by heavy and continuous air currents at different elevations.

7.3.12 Rammed Earth Brick

The need for ecological sustainability has inspired the revival of an ancient building technique: the rammed earth brick. This method employs raw materials that are obtained in an environmentally responsible manner. Construction of a rammed earth building has become more streamlined with the help of modern technology, but the traditional methods of the construction stage are still used. Stabilizing components like compressed earth or concrete is used with hard materials like clay or gravel and a damp earth mixture to make sturdy, solid walls. Rammed earth bricks are perfect for sustainable construction due to their temperature-regulating properties and low environmental effect throughout their production and procurement processes. In addition to reducing the number of secondary emissions a building produces, rammed earth construction benefits in winter as well as summer.

7.4 ECO-FRIENDLY BUILDING POWER SAVINGS

Sustainable building, green building, and green construction are all synonymous with one another and may be used interchangeably to describe the same thing. To guarantee the building will last for its intended purpose and for future generations to come, sustainable construction practices must be included throughout the design and construction phases. Using the least number of resources possible in its construction, operation, repair, demolition, and upkeep are all key components of a sustainable building. The process necessitates close coordination between the architect, client, and building engineers throughout the duration of the building's development. The purpose is to ensure that the methods used in construction and building have minimal, long-lasting, and economically beneficial effects on humans and the environment. Other goals include the efficient use of resources and energy, improved vocational health, water preservation, waste reduction, and pollution prevention. By conserving energy, the following measures may be taken to lessen the burden on the building's power supply and, in turn, lessen its impact on the environment by lowering CO_2 emissions.

Daylighting design principles allow for less reliance on artificial lighting throughout the day, lowering the building's overall energy footprint (Manikandan et al., 2022). Using locally sourced materials and minimizing the building's reliance on main energy sources are also crucial for reducing carbon emissions. Third, while designing for operation, it is important to use less energy than was originally generated using fossil fuels. When a building generates all its electricity from renewable resources, it is said to be zero-emission or very energy efficient. The financial benefits of renewable energy systems are outlined in contrast to more traditional energy options.

For long-term viability, it's important to maximize energy conservation. The composition of a passive solar architecture design may aid in energy saving since building design is directly tied to electricity consumption. Using renewable sources of energy such as the sun's rays for free heating, daylighting, and cooling is the goal of passive solar architecture. The result is a pleasant interior with less need for electricity from other sources. To achieve net-zero energy use, passive solar design concepts may be applied to both new constructions and to existing structures of varying architectural tastes. By incorporating several passive solar design elements into a building, mechanical heating and air-conditioning systems, as well as artificial lighting during the day, may be eliminated or greatly reduced. To save costs, architects and builders pay special attention to solar orientation. It is not necessary to make the building's design very complicated, but it should consider window technology, local temperature, and solar geometrical. Any structure may benefit from passive solar design provided it is situated in an appropriate location.

Energy fluxes associated with natural convective, conduction, and radiation provide the basis of passive solar energy's fundamental natural processes. A building's material determines whether it will transmit, reflect, or absorb the sun's rays as they hit the building. As a bonus, the air moves due to the sun's heat, and this motion may be predicted in man-made environments. The basic reactions to solar heat determine where the home's cooling and heating systems are located, what materials are used, and how the house is laid out. The sun's rays are harnessed without the need of any moving parts in passive solar-powered systems. Smart building energy management is possible via the use of nearly zero-energy building principles.

In order to use the sun's energy when it is not at its peak, a passive solar heating system stores it in a material and releases it at night or during cloudy days. It's possible that the sun's heat will keep everyone inside comfortable while other parts of the structure or its materials soak up the weather for later use. Energy-efficient door and window frames, insulation, natural lighting, ventilating, and shading all work together to keep a building at a reasonable temperature without wasting too much power. The method incorporates a thermal chimney, opening windows, and wing walls. Convection causes warm air to ascend to the building's higher levels, where it may be released to the outside via vents installed there. The lower-level vents might bring in colder air while this is happening. To use the cooler air outside, trees are often placed next to the structure, and their branches act as a natural vent for the lower or bottom vent.

7.5 BUILDINGS THAT COMBINE ENERGY-SAVING AND ECO-FRIENDLY TECHNOLOGIES TO RAISE THE STANDARD FOR GREEN DESIGN

The term "renewable energy" refers to the power that is generated by means that will perpetually replenish itself. The sun's rays and the Earth's internal heat may be harnessed to create various renewable energies. According to Singh et al. (2022) and Juyal et al. (2020a), renewable energy along with the technologies such as Cloud Computing, Internet of Things, and Wireless Networks has paved a way for smarter mechanisem for non-depletable energy sources like the wind, sun, ocean, and the water that is used to generate electricity and heat. Wind power, solar power, hydroelectric power, and many other forms of technology may all be derived from renewable energy sources. According to the 2007 global status report, 18% of the world's energy consumption in 2006 came from renewable, non-depletable resources, with 13% of it came from organic matter by-products like wood fires.

Hydroelectricity provided 3%, making it the second most abundant renewable resource. Around 0.8% of the world's energy comes from cutting-edge sources including the sun, the wind, the ocean, and the Earth's magma. Since no other widely accessible sources compare, the scientific possibilities for their application are enormous. The widespread use of thermal/solar photoelectric and collecting cells might efficiently handle a great quantity of electricity and heat demands of diverse structures. In the future, renewable energy sources including water, biomass, wind, and maybe even more renewable energy sources will be used to reduce dependency on traditional energy sources. Some possible answers to the issue of greenhouse gas emissions include nuclear power and renewable energy sources. Environmentally friendly, risk-free, and biosphere-harmonizing renewable energy sources are widely diffused and the preferred alternative these days. Because they are always replenishable, people may use them with little investment or faith in the market.

More jobs, sustainability, a longer lifetime for energy supplies, and a constant and consistent energy supply are just a few of the benefits that come from using renewable energy sources to propel technological developments. Despite the high cost, solar energy systems are designed to comply with international and European regulations. The only plausible answer is that solar technology can be employed in biosphere conditions without compromising the quality of today's built environment. Since the cost of producing energy from finite resources like nuclear power, natural gas, coal, and oil is rising, any innovation that might lessen or eliminate reliance on these resources would be a benefit to national economies. Solar energy systems and implementation may be mounted on buildings to provide passive cooling, lighting, heating, and electricity protection. Flat or sloping roofs, as well as the facade, are all suitable locations for installing photovoltaic cells and solar thermal collectors. Buildings constructed in accordance with sustainable building architectural styles can reduce their energy needs and their impact on the ecosystem by making use of new special glasses and heat-enclosing materials, such as different applications, that effectively minimize the amount of heat lost in the winter and power resource usage for cooling in the summer. More than half of the current energy used to create prototype buildings might be saved in the future if this trend continues. Several factors, including cost, aesthetics, design, and environmental compatibility, go into the choice to install various solar energy units and devices. More and more people are turning to solar-powered systems as a practical and aesthetic alternative to diesel engines because of their reduced environmental impact and the fact that they can be used in harmony with the existing natural and local settings of the ecosystem with some forethought and in-depth study of the area's ecology. An increase in the number of solar power plants might inspire new bioclimatic building designs that cut down on energy use. The study found that the construction industry accounted for around 35% of overall energy consumption and 40% of gaseous emissions. Solar energy-based cooling and warming systems have been shown to have an around 60% stored proportion of energy. The European Commission has released a new directive for new buildings that requires higher standards for energy efficiency. Because of this, it is clear that employing renewable energy sources to create attractive and useful structures may aid in alleviating poverty.

Multiple worldwide solar integration systems are used to power eco-friendly structures. For single-family homes, such systems might be installed as standalone modules in a series configuration, with a separate insulated storage tank located within the structure. In the case of big applications, this is inevitable. Multicoloured absorbers in solar collectors, instead of only black ones, provide a promising alternative for many solar energy uses. The thermal efficiency of coloured collectors is lower than that of conventional black collectors because of the poorer absorption of the former. Costing an additional 20% to expand the collector's area, these collectors are not as thermally efficient as their counterparts with a black absorber, but they are otherwise equivalent. A greater reliance on solar thermal collectors may be possible with the installation of blue collectors in white island structures and reddish-brown collectors on slanted roofs.

Due to the prevalence of asphalt shingles in the construction industry, solar collectors may be laid out in rows of uniform length, with adequate clearance between them to avoid winter shading. Intermediate solar irradiance on the special edition aperture surface could be delivered if the booster reflections were positioned at the base of the second row from the collecting head at the top of a row. Installing this sort of gear in the summer through fall is perfect for the collector's good performance in hot areas that need cooling modifications, as the windows may aid enhance thermal energy output by 20%–50%. The most efficient technique for reducing the intensity of scattered sunlight is the employment of glass coatings and visual difficulties faced by the light's reflection. As a result, unglazed photovoltaic systems are well-suited for use in lower temperature-based activities, such as indoor pools, where glare is less of a problem. Unglazed solar thermal collectors, which are less common than glazed solar collectors, may be used to produce steam to 35 degrees Celsius. It is possible that the exposed collection might be combined with the coloured surface to provide a more original integration of photovoltaic panels on the slanted roofs and entrances of a wide variety of buildings.

7.5.1 Systems That Combine Photovoltaic and Thermal (PV/T) Energy

Concentrated solar photovoltaic/thermal (PV/T) solutions are the way to go for both heating and power. These setups are designed to transform solar energy into useful electricity as efficiently as possible. Accordingly, they are constructed from solar modules coupled to machinery for extracting heat from water or air. Air-conditioning might be improved with the installation of such equipment.

7.5.2 Construction of Atriums Using Fresnel Lenses

Optical instruments such as Fresnel lenses are used to focus solar energy. When compared to standard thick lenses, these thinner ones have a shorter focal length, take up less space, and are less expensive. Frameless lenses are better suited for lighting management of the interior space of the structure due to their ability to separate the direct sun radiating from the dispersed sun radiations, thus producing light with more uniform intensity and fewer shadows. For the most comfortable indoor temperatures and light levels, this Fresnel lens-based system is ideal.

7.5.3 Integration of Wind/Solar Systems

The integration of wind and solar energy systems into a single building is another cutting-edge topic of study. These two methods are unique and interesting among renewable energy options for the built environment. Converting sunlight into electricity systems, which include thermal collections or photovoltaics for collecting heat and power requirements, may be installed or operated on the horizontal or sloping roofs, facades, and other surfaces of houses, hostels, and buildings of many other types. It is possible to place small-scale wind turbines (WTs) on the tops of buildings in locations with enough wind speed. When used in mesh-connected big housing neighbourhoods, hotels, etc., compact turbines are thought to provide a substantial electricity supply and are also well-suited for distributed systems.

Furthermore, heat collectors and solar cells may be added into their design. As a result, several other types of devices have emerged, including wind solar collectors, which focus the wind in a narrow beam, and wind generators, which spin on either a horizontal or vertical axis. Energy production from solar and wind may get significant assistance from hybrid diesel systems or photovoltaic–wind systems. A combination of wind turbines and solar cell systems is effective for most days and even nights in regions with abundant sunlight and wind, such as the Greek islands. Truth be told, this lasts for a good amount of the year. Therefore, we need thermal collectors that can meet our demands throughout the year. The integration of photovoltaic (electricity and thermal) or wind (electronic) systems is an entirely new idea that has been established. Most people agree that they are best suited for use in rural areas where power is provided by standalone generators or small-scale grids. Grid-connected projects may make use of PV/T or WT-based technologies. Since wind and solar power alone can't meet the electrical needs of these isolated systems, diesel generators must be brought forward.

7.5.4 The Financial Impact of Renewable Energy Technologies

The cost per unit of electricity generated by a WT system is affected by the system's utilization rate and initial investment. Alternately, experts and industry insiders would argue that better aerofoils, bigger turbines, and the adoption of more productive operating practices can all help bring down the cost of wind generation. Capital costs may also go down as a result of technological developments that allow for the use of lighter design materials, which in turn reduce production costs. Many time-tested techniques are used to collect the required thermal energy. Variations in solar panel cost have less of an impact on the unit price of thermal energy delivered by a solar power system than do changes in the unit cost of gasoline and the effectiveness with which fuel is utilized to heat water. It has been discovered that the unit price of consuming thermal electricity produced is significantly affected by the number of total days on which the investor uses or gets water heated by solar power. If you're looking to save money on your water heating bills, installing a solar water heater is a good option since its per-unit price is cheaper than that of more traditional methods. To provide purified water in far-flung locations with little precipitation and limited freshwater supplies, the simplest technology to yet have been used is the tried-and-true solar kind. It is more challenging to distil using traditional methods that rely on fossil fuels than it is to obtain specified amounts using solar energy. The costs of running, fixing, and amortizing the old systems are significant. The unit price of distilled water produced using the traditional distillation process is influenced by the aforementioned elements as well as the cost of the fuel used in its production. Using solar energy to power a distillation system adds complexity to plant management compared to typical plants, although the systems' core functionality remains unchanged from traditional desalination processes, with the exception of their information subsystems and power-generation mechanisms. Additionally, renewable sources such as wind, biomass, ocean, and thermal energy may be used in liquid–liquid extraction.

7.6 RENEWABLE ENERGY TECHNOLOGIES AND THE LONG-TERM VIABILITY OF ECO-FRIENDLY BUILDINGS

Common and reliable renewable energy-producing choices for eco-friendly, ecological, and green buildings include wind, hydropower (for rural regions along a river), geothermal (heat of the Earth), and the sun. Using renewable resources like these is critical when it comes to green building and lowering our carbon footprint. The availability of these resources has also permitted the introduction of several initiatives meant to cut down on energy waste. Cleaner energy technology may provide a bigger share of a building's gross energy demand at the same price as a less efficient construction. The size of wind generators has increased by a factor of ten thanks to rapid technological advancements. The units' output has increased from 50 kW to 5 MW in the last 25 years, and their cost has decreased by more than half in that time. With the virtue of smart networks, the sustaianable development goal 7 of Clean and Green Energy is achievable (Juyal et al., 2020b; Juyal et al., 2018) Photovoltaic-based systems of energy production are one of the rapidly growing areas in the energy sector, and try to increase energy (35%) more than the last decades.

7.7 CONCLUSION

It's crucial to consider the fact that green construction features often come at a premium. As a result, the cost of constructing environmentally friendly, risk-free buildings rises inexorably, as these structures need the use of high-efficiency, high-quality materials, and a more involved construction procedure. Researchers, scientists, architects, designers, etc., are now keeping an eye on safe, sustainable, eco-friendly development activities that may help others discover other approaches that would deliver global and societal advantages at no extra cost. Construction of environmentally friendly structures reduces financial outlay by using renewable energy methods. By using renewable energy technologies in building construction, we can lessen our reliance on finite fossil fuels and lessen the damage we do to the environment when we burn up our supplies of raw materials and electricity. In conclusion, using renewable energy technology for green buildings is a safe endeavour that helps move the biosphere closer to a more secure and sustainable future.

REFERENCES

Ahmad, S., Mishra, S. and Sharma, V. (2023). Green Computing for Sustainable Future Technologies and Its Applications. In Grima, S., Sood, K. and Özen, E. (Eds.), *Contemporary Studies of Risks in Emerging Technology, Part A (Emerald Studies in Finance, Insurance, and Risk Management)*, Emerald Publishing Limited, Bingley, 241–256. https://doi.org/10.1108/978-1-80455-562-020231016

Alena, B. and Libor, G. (2012). Green ICT Adoption Survey Focused on ICT Life-cycle from the Consumer's Perspective (SMEs). *Journal of Competitiveness*, *4*(4), 109–122.

Ali, M.A., Balamurugan, B., Dhanaraj, R.K. and Sharma, V. (2022). IoT and Blockchain based Smart Agriculture Monitoring and Intelligence Security System. In *2022 3rd International Conference on Computation, Automation and Knowledge Management (ICCAKM)*, Dubai, United Arab Emirates, 1–7. https://doi.org/10.1109/ICCAKM54721.2022.9990243.

Apulu, I. and Latham, A. (2009). Information and Communication Technology Adoption: Challenges for Nigerian SMEs. *TMC Academic Journal*, *4*(2), 64–80.

Boughton, C. (2009). What is an ICT Professional Anyway? *Australasian Journal of Information Systems*, *16*(1), 149–163.

Davis, F.D. (1989). Perceived Usefulness, Perceived Ease of Use, and User Acceptance of Information Technology. *Management Information Systems Quarterly*, *13*(3), 319–339.

Dedrick, J. (2010). Green IS: Concepts and Issues for Information Systems Research. *Information Systems Management*, *25*(2), 159–173.

Dhatterwal, J.S. and Kaswan, K.S. (2021). Intelligent Agent Based Case Base Reasoning Systems Build Knowledge Representation in COVID-19 Analysis of Recovery of Infectious Patients. In Sacfi Nandan Mohanty, Shailendra K. Saxena, Suneeta Satpathy and Jyotir Moy Chatterjee (Eds.), *Applications of Artificial Intelligence in COVID-19*, Springer, Singapore, 185–209.

Dhatterwal, J.S., Dixit, S. and Srinivasan, S. (2020). Implementation of Case Base Reasoning System using Multi-Agent System Technology for a Buyer and Seller Negotiation System. *International Journal of Electronics and Communication Engineering*, *7*(3), 63–67.

Dhatterwal, J.S., Kaswan, K.S. and Pandey, A. (2022). Implementation and Deployment of 5G-Drone Setups. In Arun Solanki, Sandhya Tarar, Simar Preet Singh Akash Tatal (Ed). *The Internet of Drones*, Apple Academic Press, Greater Noida, India, 47–64.

Durgalakshmi, K., Anbarasu, P., Karpagam, V., Venkatesh, A., Kannapiran, B. and Sharma, V. (2022). Utilization of Reduced Switch Components with Different Topologies in Multi-Level Inverter for Renewable Energy Applications-A Detailed Review. In *2022 5th International Conference on Contemporary Computing and Informatics (IC3I)*, Uttar Pradesh, India, 913–920. https://doi.org/10.1109/IC3I56241.2022.10073430.

Hart, S.L. (1997). Beyond Greening: Strategies for a Sustainable World. *Harvard Business Review*, *75*(1), 66–76.

Juyal, V., Saggar, R. and Pandey, N. (2018). Clique-Based Socially-Aware Intelligent Message Forwarding Scheme for Delay Tolerant Network. *International Journal of Communication Networks and Distributed Systems*, *21*(4), 547–559.

Juyal, V., Saggar, R. and Pandey, N. (2020a). On Exploiting Dynamic Trusted Routing Scheme in Delay Tolerant Networks. *Wireless Personal Communications*, *112*, 1705–1718.

Juyal, V., Saggar, R. and Pandey, N. (2020b). An Optimized Trusted-Cluster–Based Routing in Disruption-Tolerant Network Using Experiential Learning Model. *International Journal of Communication Systems*, *33*(1), e4196.

Kounatze, C.R. and Vickery, G. (2009). *Towards Green ICT Strategies: Assessing Policies and Programmes on ICT and the Environment.* OECD Report.

Manikandan, N., Ruby, D., Murali, S. and Sharma, V. (2022). Performance Analysis of DGA-Driven Botnets Using Artificial Neural Networks. In *2022 10th International Conference on Reliability, Infocom Technologies and Optimization (Trends and Future Directions) (ICRITO)*, 1–6. https://doi.org/10.1109/ICRITO56286.2022.9965044.

Mariani, M. and Imam, K. (2012). A Preliminary Study of Green IT Readiness in Indonesian Organizations. *Journal of Energy Technologies and Policy*, *2*(5), 1–10.

Molla, A., Cooper, V.A. and Pittayachawan, S. (2009). IT and Eco-sustainability: Developing and Validating a Green IT Readiness Model. In *Proceedings of the 13th International Conference on Information Systems (ICIS)*, USA, Christchurch, New Zealand, 141.

Molla, A. and Cooper, V.A. (2010). Green IT Readiness: A Framework and Preliminary Proof of Concept. *Australasian Journal of Information Systems*, *16*(2), 5–23.

Ogunyemi, A.A. and Johnston, K.A. (2012). Exploring the Roles of People, Governance and Technology in Organizational Readiness for Emerging Technologies. *The African Journal of Information Systems*, *4*(3), 100–119.

Page, G.A. and Hill, M. (2008). Information, Communication, and Educational Technologies in Rural Alaska. *New Directions for Adult and Continuing Education*, *117*, 59–70.

Philipson, G. (2010). A Green ICT Framework: Understanding and Measuring Green ICT. In *Connection Research Services Pty Ltd* (ABN 47 092 657 513), Australia.

Pushpalatha, N., Jabeera, S., Hemalatha, N., Sharma, V., Balusamy, B. and Yuvaraj, R. (2022). A Succinct Summary of the Solar MPPT Utilizing a Diverse Optimizing Compiler. In *2022 5th International Conference on Contemporary Computing and Informatics (IC3I)*, Uttar Pradesh, India, 1177–1181. https://doi.org/10.1109/IC3I56241.2022.10072844.

Raj, A., Sharma, V. and Shanu, A.K. (2022). Comparative Analysis of Security and Privacy Technique for Federated Learning in IOT Based Devices. In *2022 3rd International Conference on Computation, Automation and Knowledge Management (ICCAKM)*, Dubai, United Arab Emirates, 1–5. https://doi.org/10.1109/ICCAKM54721.2022.9990152.

Singh, A., Dhanaraj, R.K., Ali, M.A., Balusamy, B. and Sharma, V. (2022). Blockchain Technology in Biometric Database System. In *2022 3rd International Conference on Computation, Automation and Knowledge Management (ICCAKM)*, Dubai, United Arab Emirates, 1–6. https://doi.org/10.1109/ICCAKM54721.2022.9990133.

Venkatesh, V. and Davis, F.D. (2000). A Theoretical Extension of the Technology Acceptance Model: Four Longitudinal Field Studies. *Management Science*, *46*(2), 186–204.

Wabwoba, F., Wanyembi, G.W., Omuterema, S. and Mutua, S.M. (2013). Pervasiveness of Green ICT Awareness Amongst Kenyan ICT Personnel. *International Journal of Application or Innovation in Engineering and Management*, 2(1), 93–104.

8 Strategies for Sustainable Practices in a Post-Pandemic World

Jagjit Singh Dhatterwal, Kuldeep Singh Kaswan, Sanjay Kumar, Vandana Sharma, and Balamurugan Balusamy

CONTENTS

8.1 Introduction 149
8.2 COVID-19 Impacts 150
8.3 Improve Employee Morale and Output, Give Them More Autonomy in Their Work 153
8.4 Embrace Green Capitalism 154
8.5 The Rise of Indian Education Technology 154
8.5.1 COVID-19 Impact on Indian Education System 155
8.5.2 Overview of the Indian Edtech Industry 155
8.5.3 Future of the Edtech Sector 155
8.5.4 Blended Online Mentoring Programmes 156
8.5.5 Prospects for the Future 157
8.5.6 Edtech Companies as a Sustainable Business Model 157
8.6 Conclusion 158
References 158

8.1 INTRODUCTION

It is important to fully recognize COVID-19's far-reaching effects. Almost 4 million individuals had contracted the illness, and another 270,000 had died. Extremely high death tolls reveal the disaster's devastating scope. Although COVID-19 has proven to be a threat to people of all ages, including the physically active, it seems to have a more pronounced impact on the elderly and those with pre-existing health conditions (Adelodum, 2020). The risk that persons in substandard or disadvantaged sectors of society would be incautious is increased by factors like cooperation in preparedness and insufficient community care webs. Therefore, the pandemic may contribute to the widening of already existing economic gaps. Both healthcare and political movements are strong foes, with the latter being ill-equipped to pull off a feat of this calibre (AI-Youbi, 2020). A worse financial disaster than the monetary

DOI: 10.1201/9781003348313-8

crisis of 2008 compounds these demands on social stability and administrative burdens. Profiting from the epidemic's diverse perspectives, peoples and cultures in environmental stress, and global movement will need real-world interpretations and recovery procedures.

Further evidence that future superbugs like COVID-19 are more probable may be found in studies of historical outbreaks. COVID-19 is the third reported coronavirus epidemic after SARS in 2003 and MERS in 2012. Considering that the present period, the Anthropocene, has already seen ignorance due to climate change and the elimination of biodiversity, the disturbing pattern of three pandemics in 20 years becomes even more dramatic (anywhere humans have overpopulated areas of the Earth). This indicates the biosphere is more susceptible to catastrophes of similar or higher magnitude. This is the setting in which the "human persuaded issue of international and inter co-evolutionary links" has been cited as a possible cause for the increased likelihood and severity of epidemics like COVID-19 (Alibegovic, 2020). Some researchers believe that the fast and unrestricted travel of sick persons made feasible by modern transportation contributed to the COVID-19 pandemic. It's possible that these hypothetical etiological elements are making an appearance. In order to address the fundamental challenges posed by this pandemic, it is urgent that we drastically cut back on unchecked land expansion and the ecological destruction it brings, and that we fully implement climate change countermeasures to lessen the extensive impacts of climate change on networks. It is also crucial to build sturdy zones that can cooperate with one another so that dangers may be mitigated as soon as possible (Allen, 2018). Both the General Assembly and the Sustainable Development Agenda 2030 have served as foundational documents for long-term flexibility and viability. Encouraging the development of regional circulating and ecological spheres (Regional-CES), which seek to achieve holistic sustainable development at the regional level by combining efforts to attain various, communal, financial, and environmental boards are a practical means of progressing in the long run. Since the majority of affected persons tend to congregate in urban areas, it stands to reason that that is where the virus will spread quickest. Cities and other characteristics, such as the prevalence of illness hotspots, may be measured in a number of ways (Amankwah-Amoah, 2020).

8.2 COVID-19 IMPACTS

In the first place, on a physical level, people are okay with carrying out essential daily activities like travelling and spending money in rather overcrowded areas. Sometimes the urban poor have no choice but to live in filthy, overcrowded conditions.

Second, economic expansion is propelled by major cities, making possible higher standards of living and the emergence of new professions. Unsustainable manufacturing, rather than consumable adornment, is still linked to economic expansion. Megacities reject all types of ecological responsibilities, including those related to the housing, food, clothing, and transportation of their massive populations (Kaswan, 2021).

Third, the health of city dwellers was already deteriorating before pollution was brought to the environment. For quite some time now, research has linked urban air pollution to an increased death toll. Once again, COVID-19 has shown the vulnerability of global enterprises to global hazards, particularly those with extensive resource constraints and a dependence on foreign travel. An increasing worldwide hazard is being caused by climate change, habitat destruction, and other concerns for sustainability. The 2008 food price crisis and the 2011 floods in Bangkok are two examples. The COVID-19 issue has disrupted the car industry's critical distribution network. In order to safeguard the international supply chain, the textile industry has, for the time being, jeopardized the livelihoods of low-income workers. It has been claimed, for instance, that despite the nationwide lockdown in Bangladesh, 200,000 garment workers were forced to return to their jobs in sweatshops.

The global tourism industry has also been hit hard by the COVID-19 epidemic. At many popular tourist destinations across the globe, overcrowding, trash, and disruptions to native traditions have long been major issues due to an increase in international visitors. The current concern is that travel unintentionally aided in the propagation of the infection. Vulnerabilities in such global systems must be taken very seriously, with duc weight given to the advantages and hazards involved. The aforementioned issues need action on a worldwide scale. In light of the fact that the problem will have a disproportionate impact on disadvantaged communities in developing nations, efforts to solve it need to be ramped up (Ashford, 2020).

The COVID pandemic is also evidence that the rising globalization of risk may have catastrophic social and economic consequences, including as the breakdown of global supply networks and travel. Therefore, it is believed that local and international policies and initiatives are required to strengthen supply networks' sustainability and resilience. Additionally, there has to be a major improvement in regulations and procedures to encourage sustainable tourism all over the globe. It's also possible to consider alternative options, such as whether an "Earth Solidarity Tax" is even possible. Right now, several nations have implemented a solidarity tax on plane tickets, but it is not widespread. The funds generated by these programmes should be used to bolster the lives of the most vulnerable citizens, especially in underdeveloped nations (Bansal, 2019). IGES might collaborate with local and foreign partners to do research on better global governance to effectively manage globalized threats, bearing the aforementioned considerations in mind. To assist in creating a more resilient and environmentally friendly world and to reduce future catastrophic risks, the Institute for Global Environmental Strategies (IGES) has initiated and will continue to support experiments on the consequences of COVID-19 and suggestions for effective collective security and multilayered recovery and reaction in the short, medium, and long terms. IGES will keep working with regional and global allies to support transformative change that benefits human health and the global environment (Dhatterwal, 2020).

The COVID-19 epidemic triggered a worldwide economic downturn and the collapse of several businesses. Therefore, it is predictable to know going down alterations in sustainable plans before scheduled resources available, ensuring strength, competence, and ecological protection. The significant slowdown in financial activities caused by the epidemic negatively impacts sustainable schemes. As a result, it represents a serious threat to the development of the economy in the long term. To keep these unwelcomed by-products of the pandemic from threatening to derail otherwise sustainable progress, it may be necessary to adopt some novel financial and commercial measures. Creative solutions (Ahmad et al., 2023) can take many forms, including new approaches to channelling funds, green infrastructure planning, taxing carbon productions (at the local or global equivalent), streamlining the issuing of green commitments, establishing a value system for green praise total score, and lessening the risks associated with green reserves. Humans are a very resilient species. In the face of wars, tsunamis, and economic booms, we dust ourselves off and turn the media on. Now we can leave COVID-19 behind us (Barcaccia, 2020). But is it too much to hope that the very fact of human presence may contribute to a healthier ecosystem, given how pervasive the sickness seems to be?

Despite all that has occurred, we still have cause for optimism. One study by the Centre for Research on Energy and Clean Air found that in April 2018, there were 11,000 fewer deaths in Europe attributable to air pollution than there were in April 2017. If similar measures are done in other large polluted locations in Asia and across the globe, it is feasible that many lives might be saved during the pandemic (Barreiro, 2020).

But we know from past experience that things don't improve by themselves. We need to do something. So, what do sustainability experts from the world's business schools recommend firms do to make a better society a reality as the crisis abates?

Professor of Information Systems for Sustainability and Director of the Institute of Energy Economics at the University of Cologne, Wolf Ketter, is convinced that the COVID-19 pandemic is a classic case of "demand congestion", in which an influx of patients overwhelms hospital resources and distorts doctors' ability to treat their patients. Another reason for its naming is that it will save health organizations time and effort in the long run by eliminating the need for them to include interplanetary planning on their to-do lists. Major sectors like advertising will be forced to face their own tests in the future as a result of the methods used to bring this one into being. We need to adjust the curvature of the chairs differently in the power transition, he explains. All viable breadwinners must act swiftly to meet the avalanches of demand in their niche markets. A machine-to-machine economy, together with artificial intelligence to aid humans in reacting swiftly to fluctuating market conditions, would be required to facilitate such complex behaviours. Fortunately, new companies offering "energy flexibility" offered new methods to break the curve (Sharma et al, 2023). Protecting critical public services will soon need the combined efforts of artificial intelligence and advanced mechanical knowledge (Barua, 2020).

Mary Martin, head of the UN Business and Human Security Initiative and senior policy scholar at LSE IDEAS, thinks that the COVID-19 crisis offers important lessons on how to contribute to establishing countries that are commercially, socially, and environmentally healthy, secure, and resilient. However, it is vital to go beyond examining the implications on individual projects like reducing emissions or improving health care when assessing the impact of the pandemic on environmentally friendly business practices. Support for human security in all its manifestations, including as collaborations with local neighbourhoods and individuals to protect people and solve a broad variety of difficulties that influence their everyday lives, must become standard procedure in the way enterprises are managed (Bascope, 2019). In order to secure the long-term success of neighbourhoods, businesses should collaborate with local officials to enhance the quality of life there. Companies' attempts to solve local concerns at the community level will strengthen their trust with common people, and such activity will be vital in the fight for a better world and a focus on the environment.

Nicolas Béfort, associate professor at NEOMA Business School and director of the Chair in Bioeconomy and Sustainable Development, argues that the COVID-19 pandemic demonstrates the magnitude of economic impact an ecological calamity may have. Moreover, it serves as a prime example of the interconnected character of contemporary society. Despite the virus's biological origins, he says that its spread to people is due to the destruction of natural ecosystems. The frightening pace at which the pandemic spread may largely be attributed to the globalization of production. Due to the reliance on timing in just-in-time production, business models are very unstable (Bodenheimer, 2020). Increased sensitivity and decreased resilience result from rising levels of household debt, making it more likely that any prospective failure anywhere in the value chain may trigger a financial disaster. Therefore, maximal growth is assumed while making economic predictions. It is simply because European countries have transferred their greenhouse gas (GHG) emissions to Asia that they can boast about having a lesser percentage of global GHG emissions than China. Last but not least, the excessive use of air and marine travel contributes to global pollution through increasing human-to-human contact (and hence transmitting vectors) and generating GHGs. Because of this, we need to rethink our production, trade, and consumption patterns to focus on self-sufficiency (Basheer et al., 2023) rather than exponential growth (Calabrese, 2018).

8.3 IMPROVE EMPLOYEE MORALE AND OUTPUT, GIVE THEM MORE AUTONOMY IN THEIR WORK

The epidemic has made it abundantly evident how important it is to bring together as many resources as possible to develop solutions to a problem of this magnitude. According to Michael Butler, a professor of Organizational Evaluation and Development at the Aston College Of business in the United Kingdom, the private sector is the most effective. When people from different fields work together, they pool their knowledge and develop novel approaches to problems. His analysis suggests, however, that effective participation and genuine leadership are prerequisites for success. And the reaction to COVID-19 could speed up the introduction of new cooperative models. Businesses that adopt an open approach to innovation do so because they recognize there is much they do not know, particularly in times of uncertainty (Chakraborty, 2020). In order to get input on supply chain choices and planning for the future, they communicate with a broad range of individuals. However, a mix of elements, including leadership reform and an inclusive and productive culture supported by updated incentive systems and promotion pathways to entrench cooperation, is crucial to effective collaboration.

Is it possible that we've reached the point where we can no longer afford to go after every possible dollar, no matter the cost in other areas? That's the opinion of Professor Konstantinos Stathopoulos of Alliance Manchester Business School, who holds the chair in Accounting and Finance there. He argues that the COVID-19 epidemic has shown how important it is for governments to take precautions for their citizens and how much money is required to do so (Dhatterwal, 2021). Having enough money for public services in the future depends in large part on having a business tax system that is both equitable and balanced. Companies should do the right thing and pay their taxes when they're due rather than employing legions of "tax consultants" to help them avoid paying them. However, it is not sufficient to just act as a socially responsible business. There may be a need to reevaluate a central business model that has gained universal acceptance since the 1980s, he argues. Maximizing profits requires following the just-in-time operational concepts and the corresponding finance strategies. But this leaves businesses vulnerable to and unprepared for extreme market fluctuations. While most businesses can't save up enough money to weather a large storm, keeping a little extra cash on hand might help them keep their most important constituents (such workers, customers, and suppliers) from feeling the pinch for a while. This notion was at the heart of the decision to increase underwriting standards in the banking industry in the wake of the 2008 financial crisis. "However, you won't find this mindset in most other fields" (Chiang, 2020).

8.4 EMBRACE GREEN CAPITALISM

David Veredas, a professor of Finance and Sustainability at Belgium's Vlerick Business School, believes a reevaluation of the nature and goals of capitalism is warranted in the wake of the global epidemic (Rajendran et al., 2022). He argues that the original intent of capitalism was to benefit all parties involved in a business, not just the stockholders. COVID-19 enables us to circle back to the origins of capitalism. All the community's members are treated as equal stakeholders in the cooperative, making it a great model of "public participation entrepreneurship". It's possible to find some of the most successful cooperative firms in Europe. One prominent Belgian example is the financial services corporation KBC, the majority of whose shareholders are cooperatives based in the Flemish region. They were instrumental in the area's rise from a regional economy to its status as one of Europe's most prosperous and enterprising regions with a top-tier educational system.

The COVID crisis, like all others, will eventually pass. The number of corporate executives with the foresight and the guts to implement even a fraction of the recommendations made by these experts will determine whether we emerge into a better, brighter development. Let's cross our fingers and hope there are a lot of them (Chowdhury, 2021).

8.5 THE RISE OF INDIAN EDUCATION TECHNOLOGY

More than 4450 educational technology (edtech) businesses were launched in India between January 2014 and September 2019, demonstrating the country's rising interest in the information technology and education industries. With the help of over 500 million internet users and smartphone owners, digital adoption was given a much-needed push. Before the epidemic, India's edtech sector was expanding slowly due to widespread pessimism about the efficacy of online education. Teachers were quite worried that a lack of technology would be a big barrier to reskilling and retraining.

8.5.1 COVID-19 Impact on Indian Education System

The COVID epidemic has devastated many sectors in India, but it has come at a critical moment for the country's developing edtech industry. In the wake of COVID-19 lockdowns, schools, universities, and universities have increased usage of a wide range of edtech goods and services. Vedantu and Byju's (tutoring), Toppr (learning), and Unacademy (online learning platform) all saw significant growth because to the lockout. Toppr's paid user base increased by 100%, while Byju's had a 7.5 million user increase during the downtime.

An estimated 320 million students in India are impacted by the COVID-19 epidemic, which has imposed novel challenges on the country's educational infrastructure. Instructions tailored to India's higher education institutions, considering the University Grants Commission's and other pinnacle education bodies' advisories of COVID-19, a total of 1000 universities and 40,000 colleges were forced to temporarily close, sending students home and prompting administrators to explore alternatives, such as online courses. The 37.5 million students and the 1.4 million instructors in the system were affected but to varying degrees.

8.5.2 Overview of the Indian Edtech Industry

The chances of kids in a nation with a historically unfavourable ratio of teachers to pupils are low. Edtech offers a complete solution to these and other problems. There are now over 4530 edtech companies in the United States and 761 in India. KPMG predicts that by 2021, online education in India would be valued $1.96 billion.

Market participants have access to a more diverse school environment via e-learning systems due to reductions in the cost of digital communications. The business was growing at double-digit rates before the pandemic, and development and acquisitions activities have increased.

8.5.3 Future of the Edtech Sector

This paradigm shift towards edtech that began in 2021 is likely to deepen and spread across the academic community in the next year. The realization among students that traditional offline players cannot provide enough support in these times of intense competition will grow.

As of the latest current count, around 250 million Indians, or roughly 20% of the country's total population, are enrolled in some sort of formal learning. Approximately 36% of India's population is still in their youth and actively pursuing (Pushpalatha et al., 2022).

There is a huge opportunity for the education industry to grow in that respect. After COVID, the educational sector went through an online transformation similar to that of other sectors. The good news is that projections show an $11.6 billion increase in the online education sector by 2026.

8.5.4 Blended Online Mentoring Programmes

There are no additional costs for students taking the IIT-JEE, NEET-UG, GATE, or UPSC-CSE exams online. Also, they won't need to give any thought to shifting their foundation. Teaching methods provide students with scheduling options that might otherwise be unavailable in the traditional classroom setting (Anisul Huq et al. 2014).

Those preparing for competitive exams may go at their own pace and see how they stack up against their peers by analyzing quantifiable information. Online mentorship programmes will be in great demand at traditional universities due to the benefits of individualized instruction.

From 2016 to 2021, the world's self-paced learning market is projected to increase (Figure 8.1). In 2016, the worldwide market for self-paced online learning was estimated at US$46.67 billion. But forecasts expect a decrease to US$33.5 billion in 2021.

In India, Byju's edtech platform had 57 million users by 2020. Compared to its rivals Vedantu and Unacademy, both of which had at least 25 million subscribers apiece, Upgrade had less than a million users (Figure 8.2).

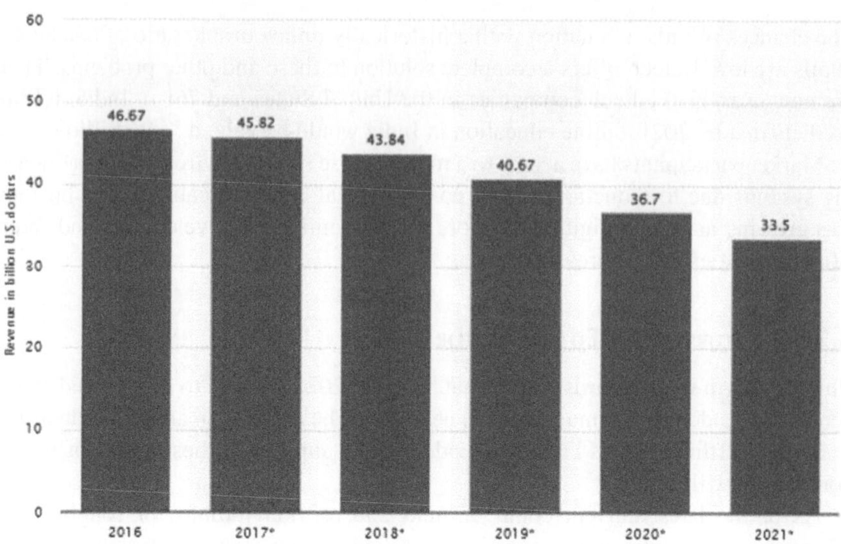

FIGURE 8.1 Money made worldwide by the self-paced e-learning market.

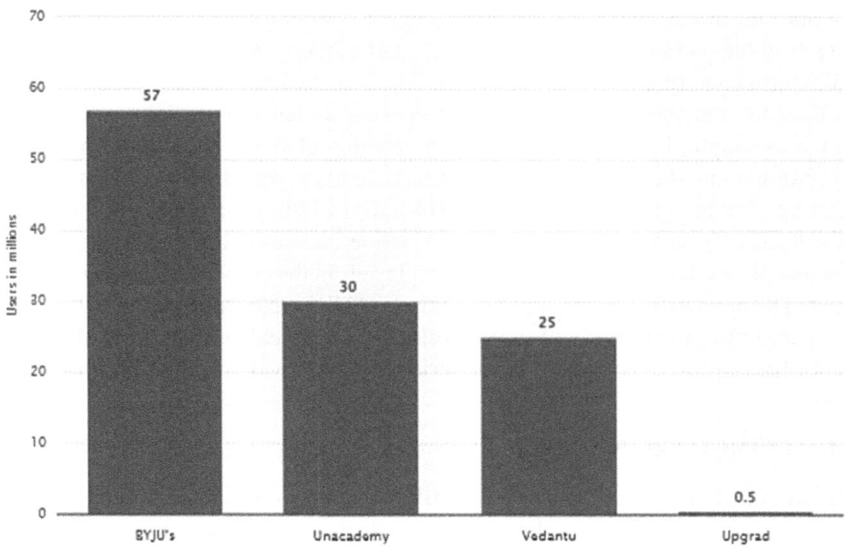

FIGURE 8.2 E-platform comparative analysis for the year 2020.

8.5.5 Prospects for the Future

In 2021, there was a major transition to using online learning resources. This trend is only going to strengthen over time. The realization among students that traditional offline players cannot provide enough support in these times of intense competition will grow. Traditional players will no longer benefit as much from high name recognition as they previously did. To that end, digital learning seems to have a bright future in 2022, particularly at the college level and in the field of entrance exams.

8.5.6 Edtech Companies as a Sustainable Business Model

A younger population, rising household incomes for parents, cheaper internet connectivity, and accessible, low-cost learning equipment are the four main factors propelling India's edtech industry forward. From the current 500 million, the number of smartphone users is projected to more than quadruple by 2025. This is possible because of wide range and use of wireless networks and existing infrastructure of Internet of Things and Delay Tolerant Networks (Juyal et al., 2018, 2020).

The counselling and test-prep sectors on the periphery have felt the effects of educational start-up companies, but this trend has yet to permeate the core of the sector. The incorporation of technology into the classroom is very beneficial, but it is not a replacement for tried-and-true methods of instruction.

Enterprises in the edtech sector would do well to remember reality in the creation of their offerings. COVID-19 expedited the adoption of digital technology within the educational sector, but edtech enterprises still need to work hard to keep their stakeholders interested and involved. Constant experimentation and new product possibilities might make this a reality. Student retention may be difficult for these edtech systems after the pandemic is over, and schools and universities reopen and life returns to normal.

If edtech start-ups are serious about preparing students for the jobs of the future, they must emphasize the need of learning new digital skills. For long-term success, diversifying into the related edtech business is another possibility worth considering. Start-ups that help students by giving them access to study materials and instructors have been very successful. Massive open online courses (MOOCs) is another tool that may be used to benefit students and employees in the corporate learning environment. The moment has come for universities to collaborate strategically with edtech companies. In fact, to push on with development and realize their aims, they have somewhat improved the quality of the services they provide via their programmes.

8.6 CONCLUSION

We have reexamined and assessed the TBL framework's social, environmental, and economic elements to assess the pandemic crisis's consequences for the long-term viability of human life. Following COVID-19, a comprehensive review of the relevant literature was conducted in order to shed light on the environmental, social, and economic aspects of sustainability. Each of the three pillars of sustainability—environmental, social, and economic—as well as their intersections and the integration of all three pillars (sustainable development) were examined to determine their effects, challenges, and potential solutions using the TBL framework. Descriptive analysis of 49 research publications provided an up-to-date picture of the state of research on the TBL of sustainability in light of the COVID-19 pandemic's consequences. In addition, a thematic analysis was used to synthesis and map the implications of the pandemic outbreak on the sustainability dimensions in several topic areas, highlighting the COVID-19's effect.

REFERENCES

Adelodun, B., Ajibade, F.O., Ibrahim, R.G., Bakare, H.O., Choi, K.S., (2020). Snowballing transmission of COVID-19 (SARS-CoV-2) through wastewater: Any sustainable preventive measures to curtail the scourge in low-income countries? *Sci. Total Environ.* 742, 140680. https://doi.org/10.1016/j.scitotenv.2020.140680.

Ahmad, S., Mishra, S., Sharma, V., (2023). Green computing for sustainable future technologies and its applications. In Grima, S., Sood, K., Özen, E. (Eds.), *Contemporary Studies of Risks in Emerging Technology, Part A (Emerald Studies in Finance, Insurance, and Risk Management)*, Emerald Publishing Limited, Bingley, 241–256. https://doi.org/10.1108/978-1-80455-562-020231016

Alibegovic, M., Cavalli, L., Lizzi, G., Romani, I., Vergalli, S., (2020). COVID-19 & SDGs: Does the current pandemic have an impact on the 17 Sustainable Development Goals? A qualitative analysis, FEEM BRIEF. https://doi.org/https://www.feem.it/m/publications_pages/brief07-2020.pdf.

Allen, C., Metternicht, G., Wiedmann, T., (2018). Initial progress in implementing the Sustainable Development Goals (SDGs): A review of evidence from countries. *Sustain. Sci.* 13, 1453–1467. https://doi.org/10.1007/s11625-018-0572-3.

Al-Youbi, A.O., Al-Hayani, A., Bardesi, H.J., Basheri, M., Lytras, M.D., Aljohani, N.R., (2020). The King Abdulaziz University (KAU) pandemic framework: A methodological approach to leverage social media for the sustainable management of higher education in crisis. *Sustainability* 12, 4367. https://doi.org/10.3390/su12114367.

Amankwah-Amoah, J., (2020). Stepping up and stepping out of COVID-19: New challenges for environmental sustainability policies in the global airline industry. *J. Clean. Prod.* 271, 123000. https://doi.org/10.1016/j.jclepro.2020.123000.

Anisul Huq, F., Stevenson, M., Zorzini, M., (2014). Social sustainability in developing country suppliers. *Int. J. Oper. Prod. Manag.* 34, 610–638. https://doi.org/10.1108/IJOPM-10-2012-0467.

Ashford, N.A., Hall, R.P., Arango-Quiroga, J., Metaxas, K.A., Showalter, A.L., (2020). Addressing inequality: The first step beyond COVID-19 and towards sustainability. *Sustainability* 12, 5404. https://doi.org/10.3390/su12135404.

Bansal, S., Garg, I., Sharma, G., (2019). Social entrepreneurship as a path for social change and driver of sustainable development: A systematic review and research agenda. *Sustainability* 11, 1091. https://doi.org/10.3390/su11041091.

Barcaccia, G., D'Agostino, V., Zotti, A., Cozzi, B., (2020). Impact of the SARS-CoV-2 on the Italian Agri-Food sector: An analysis of the quarter of pandemic lockdown and clues for a socio-economic and territorial restart. *Sustainability* 12, 5651. https://doi.org/10.3390/su12145651.

Barreiro-Gen, M., Lozano, R., Zafar, A., (2020). Changes in sustainability priorities in organisations due to the COVID-19 outbreak: Averting environmental rebound effects on society. *Sustainability* 12, 5031. https://doi.org/10.3390/su12125031.

Barua, S., (2020). Financing sustainable development goals: A review of challenges and mitigation strategies. *Bus. Strateg. Dev.* 3, 277–293. https://doi.org/10.1002/bsd2.94.

Bascopé, M., Perasso, P., Reiss, K., (2019). Systematic review of education for sustainable development at an early stage: Cornerstones and pedagogical approaches for teacher professional development. *Sustainability* 11, 719. https://doi.org/10.3390/su11030719.

Basheer, S., Singh, K. U., Sharma, V., Bhatia, S., Pande, N., & Kumar, A. (2023). A robust NIfTI image authentication framework to ensure reliable and safe diagnosis. *PeerJ Comput. Sci.* 9, e1323.

Bodenheimer, M., Leidenberger, J., (2020). COVID-19 as a window of opportunity for sustainability transitions? Narratives and communication strategies beyond the pandemic. *Sustain. Sci. Pract. Policy* 16, 61–66. https://doi.org/10.1080/15487733.2020.1766318.

Calabrese, A., Castaldi, C., Forte, G., Levialdi, N.G., (2018). Sustainability-oriented service innovation: An emerging research field. *J. Clean. Prod.* 193, 533–548. https://doi.org/10.1016/j.jclepro.2018.05.073.

Chakraborty, I., Maity, P., (2020). COVID-19 outbreak: Migration, effects on society, global environment and prevention. *Sci. Total Environ.* 728, 138882. https://doi.org/10.1016/j.scitotenv.2020.138882.

Chiang, J., Hanna, A., Lebowitz, D., Ganti, L., (2020). Elastomeric respirators are safer and more sustainable alternatives to disposable N95 masks during the coronavirus outbreak. *Int. J. Emerg. Med.* 13, 1–5. https://doi.org/10.1186/s12245-020-00296-8.

Chowdhury, P., Paul, S.K., Kaisar, S., Moktadir, M.A., (2021). COVID-19 pandemic related supply chain studies: A systematic review. *Transp. Res. Part E Logist. Transp. Rev.* 148, 102271. https://doi.org/10.1016/j.tre.2021.102271.

Dhatterwal, J.S., Kaswan, K.S., (2021). Intelligent agent based case base reasoning systems build knowledge representation in COVID-19 analysis of recovery of infectious patients. In Sachi Nandan Mohanty, Shailendra K. Saxena, Suneeta Satpathy, Jyotir Moy Chatterjee (Eds.), *Applications of Artificial Intelligence in COVID-19*. Springer, Singapore, 185–209.

Dhatterwal, J.S., Dixit, S., Srinivasan, S., (2020). Implementation of case base reasoning system using multi-agent system technology for a buyer and seller negotiation system. *Int. J. Electron. Commun. Eng.* 7(3), 63–67.

Juyal, V., Saggar, R., and Pandey, N. (2018). Clique-based socially-aware intelligent message forwarding scheme for delay tolerant network. *Int. J. Commun. Netw. Distrib. Syst.* 21(4), 547–559.

Juyal, V., Saggar, R., and Pandey, N. (2020). On exploiting dynamic trusted routing scheme in delay tolerant networks. *Wireless Pers. Commun.* 112, 1705–1718.

Kaswan, K.S., Gaur, L., Dhatterwal, J.S., Kumar, R., (2021). AI-based natural language processing for the generation of meaningful information electronic health record (EHR) data. In Loveleen Gaur, Arun Solanki, Samuel Fosso Wamba, Noor Zaman Jhanjhi (Eds.), *Advanced AI Techniques and Applications in Bioinformatics.* Springer, New York, 41–86.

Pushpalatha, N., Jabeera, S., Hemalatha, N., Sharma, V., Balusamy, B., Yuvaraj, R., (2022). A succinct summary of the solar MPPT utilizing a diverse optimizing compiler. In *2022 5th International Conference on Contemporary Computing and Informatics (IC3I)*, Uttar Pradesh, India, 1177–1181. https://doi.org/10.1109/IC3I56241.2022.10072844.

Rajendran, S., Sathiyaraj Rajendran, Munish Sabharwal, Gheorghita Ghinea, Rajesh Kumar Dhanaraj, Balamurugan Balusamy., (2022). *IoT and Big Data Analytics for Smart Cities: A Global Perspective.* CRC Press.

Sharma, V., Balusamy, B., Thomas, J. J., Atlas, L. G., eds., (2023). *Data Fabric Architectures: Web-Driven Applications.* Walter de Gruyter GmbH & Co KG.

9 Strategies for Sustainable Supply Chains in a Circular Economy

Bibliometric Analysis and an Overview

Laxmi Pandit Vishwakarma and Rajesh Kr Singh

CONTENTS

9.1 Introduction 162
9.2 Research Methodology 164
9.3 Results of the Study 166
9.3.1 Descriptive Analysis 166
9.3.1.1 Annual Scientific Production 166
9.3.1.2 Most Cited Countries 166
9.3.1.3 Most Relevant Affiliations 166
9.3.1.4 Author Network 166
9.3.1.5 Countries Network 167
9.3.1.6 Journals List 167
9.3.1.7 Word Cloud 167
9.3.1.8 Thematic Analysis 169
9.3.1.9 Thematic Evolution 172
9.3.2 Strategies towards Sustainable Supply Chains in the Circular Economy 176
9.3.2.1 Collaborating among Partners 177
9.3.2.2 Enabling Circular Product Designing 178
9.3.2.3 Paying Particular Attention to the Packaging of the Products 178
9.3.2.4 Financial Incentives to Support and Make Affordable Circular Economy Practices 178
9.3.2.5 Leveraging Technologies and Data Support to Drive Circularity 179
9.3.2.6 Altering the Business Models 179
9.3.2.7 Reduce, Reuse, and Recycle Products 180
9.4 Conclusion, Limitation, and Future Scope 181
References 182

DOI: 10.1201/9781003348313-9

9.1 INTRODUCTION

In the recent past, the concept of a circular economy has received attention from many academicians, organizations, and government institutions (Merli et al., 2018). Due to the increase in pollution, climatic changes, and greenhouse emissions, increasing waste creates an alarming situation, highlighting the need for sustainable economic development. The usage of natural resources has grown three times since 1970 and is expected to grow twice again by 2050 (UNCTAD, 2022; World Resources Institute, 2021). Everything we use is made up of fossil fuels, metals, organic materials, or minerals but comes from plants and animals. If we continue to exploit natural resources, we will need extra planet resources to fulfil our current needs (World Resources Institute, 2021). Today the amalgamation of technologies like Blockchain technology (Ali et al., 2022; Singh et al., 2022) have paved a way for secure and smooth tractions across world. In today's era, achieving sustainability is essential to surviving in the long run. A recent United Nations Conference on Trade and Development highlights that "today's goods are tomorrow's resources at yesterday's resource prices", showing the importance of sustainability (World Resources Institute, 2021). The concepts of circular economy and sustainability are somewhat interrelated (Ahmad et al., 2023). The present supply chains are confronted with uncertainty, climatic changes, risks, changing government regulations, and imbalanced demand and supply across the globe. These changes simultaneously impact sustainability practices across the supply chains. Katiyar et al. (2018) reveal that sustainability focuses on the long-term survival of the supply chains. From the organization's view, sustainability is seeking cost reductions and optimizing human and natural resources (Durgalakshmi et al., 2022; Pushpalatha et al., 2022). Since each industry is different, the definition of sustainability and closed-loop supply chains differs (Mehrjerdi and Shafiee, 2021). Mardani et al. (2020) observed that balancing the three dimensions of sustainability is a practical challenge for supply chain management. Therefore, it is important to understand past works in the circular economy and sustainable supply chains. Past literature helps identify the relevant works conducted in this field and highlight the research gaps that can be addressed in the current study. Therefore, we conducted a bibliometric analysis of the past ten years, from 2012 to 20 October 2022. In this context, we developed our first research question (RQ):

RQ1: How has the circular economy's literature on sustainable supply chains evolved over the past ten years?

The circular economy concept aims to decouple value creation from waste generation and replace it with a closed-loop product life cycle (Sassanelli et al., 2019; Camacho-Otero, 2018). This concept transfers the linear production and consumption patterns into a closed-loop circular system. In the closed-loop circular system, the products, services, and materials are made to be used for longer durations. Moreover, it follows the reuse, recycle, and reduce principles (Merli et al., 2018; Goyal et al., 2016; Lieder and Rashid, 2016). Academicians and institutions propose several definitions of the circular economy. For example, the Ellen MacArthur Foundation (EMF, 2013) defines the circular economy as "an industrial economy that is restorative or regenerative by intention and design". Korhonen et al. (2018) define the concept of circular economy as follows:

> Circular economy is an economy constructed from societal production–consumption systems that maximize the service produced from the linear nature–society–nature material and energy throughput flow. This is done using cyclical materials flows, renewable energy sources and cascading-type energy flows. A successful circular economy contributes to all three dimensions of sustainable development. A circular economy limits the throughput flow to a level that nature tolerates and utilizes ecosystem cycles in economic cycles by respecting their natural reproduction rates.

Farooque et al. (2019) define the circular supply chain as "when a company reducing their impact on the environment related to their products and rethinking the processes and activities considering the product's life cycle". In another context, the concept of the closed-loop supply chain is when "the supply chain considers all sectors and partners, creates and organizes the network to recover, recycle and reuse its products" (Zhang et al., 2020; Kalverkamp et al., 2019; Son et al., 2018). A linear economy works on "take–make–dispose" concept. In contrast, a circular economy works on "reduce–reuse–recycle" concept (Esposito et al., 2020). The concept of linear economy focuses on "eco-efficiency".

In contrast, the circular economy concept works on "eco-effectivity". The concept of circular economy is old but came into light with sustainable development (Geissdoerfer et al., 2017). Studies have increased in this domain with the United Nations Sustainable Development Goals in 2015 (Esposito et al., 2020). Esposito et al. (2018) reveal that the circular economy consists of responsible production and consumption activities. Pinho Santos and Proença (2022) explore the sustainable supply chain for the automobile industry. The study suggests that reorganizing the supply chain is essential, along with rethinking the products from the beginning to attain a sustainable supply chain. Finally, Fassio et al. (2022) discuss the circular economy for food supply chains. Their study illustrates the typical supply chains of cheese, wine, bovine beef, dairy, and rice. Further, it evaluates how the enhancements of products and food waste can be conducted. Moreover, the study proposes strategies for sustainable development through implementation of Artificial Neural Networks (Maurya et. al., 2022), Internet of Things (Basheer et al., 2023) to optimally utilize the natiral resources. Soleimani (2021) considered a novel closed-loop supply chain approach in the mining industries to reduce stone wastage. To attain a sustainable approach, the Web Driven technologies framework (Sharma et al., 2023) consider minimizing energy consumption as the profit maximization function for meeting the objective function of the service level requirements. Brydges (2021) analyzes the concept of circular economy in the fashion industry, as it is one of the most wasteful consumer industries. The study examined how circular economy principles can be incorporated into the fashion industry. Li et al. (2021) analyze the prices, coordination, and collective decision-making among the retailers and manufacturers in a closed-loop supply chain. Yoo and Cheong (2021) used a game-theoretical model to understand the joint decision on refurbishing and inventories strategies in a closed-loop supply chain. The study evaluates how a third-party refurbishment can contribute to sustainable operations in a closed-loop supply chain.

Several kinds of literature discuss circular economy definitions, concepts, drivers, and the relationship between circular economy and sustainability. Transitioning from a linear to a circular economy is difficult for many industries. Organizations face many critical challenges, such as stakeholder management, financial requirements, regulatory support and guidelines, resource management, healthcare industry (Ajmani et al., 2022) and organizational commitments (Stewart and Niero, 2018). Yet, there are firms ready to adopt closed-loop supply chains and are in the transition phase to implement the circular economy in their supply chains. To implement it, organizations must have well-defined strategies with them. Past literature highlights that there is a lack of studies conducted to understand the strategies for implementing a circular economy (Fassio et al., 2022; Brydges, 2021; Esposito et al., 2018; Stewart and Niero, 2018). Our study aims to fill this gap. Therefore, this chapter discusses the strategies towards sustainable supply chains in the circular economy. In this context, we develop our second research question:

RQ2: What are the strategies towards sustainable supply chains in the circular economy?

To answer the research questions, the subsequent sections of the book chapter are organized in the following manner: Section 9.2 explains the research methodology used for the study. Section 9.3 highlights the study's results and is divided into two subsections: Section 9.3.1 presents the descriptive analysis drawn from the study's bibliometric analysis (using the biblioshiny website and the VOSviewer software). Section 9.3.2 reveals the strategies towards sustainable supply chains in the circular economy drawn from the literature review. Then, Section 9.4 provides the conclusion, limitations, and future scope of the study.

9.2 RESEARCH METHODOLOGY

We used bibliometric analysis as the methodology in this chapter to meet our research question 1. We used a literature review methodology to meet our research question 2. Several keywords in a string were searched at the Scopus database's title, abstract, and keyword sections. Initially, keywords such as "Sustainable Supply Chains", "Sustainable Supply Chain", "Circular Economy", "Closed Loop Supply Chain", and "Strategies" were searched using the "OR" and "AND" commands. During this search, a total of 246 documents were shown. We further used a few limitations, including the subject areas decision sciences, social sciences, business, management, and accounting.

Furthermore, we considered the past ten years; therefore, our study period was from 2012 to 20 October 2022. The publication stage was considered for the study, that is, final publications and articles in press. Further, we limited our analysis to English and considered the document type as articles and review articles. Finally, a sample of 106 documents was extracted from the Scopus database. Further used were the biblioshiny platform and the VOSviewer software for the study's descriptive analysis. Garfield (1979) found that bibliometric analysis is a "quantitative method effective for determining the intellectual structure of a scientific field". Through bibliometric analysis, we examined the annual scientific production, top ten most cited countries, top ten most relevant affiliations, author network, countries network, journals list, word cloud, thematic analysis, and thematic evolution (Goyal et al., 2016). The word cloud, drawn using the biblioshiny software and the VOSviewer software, will help understand the evolutionary perspective on how the circular economy has evolved over the years and gained academic and practitioner attention. We conducted a literature review to find the strategies towards sustainable supply chains in the circular economy. The flow chart of the research methodology is shown in Figure 9.1.

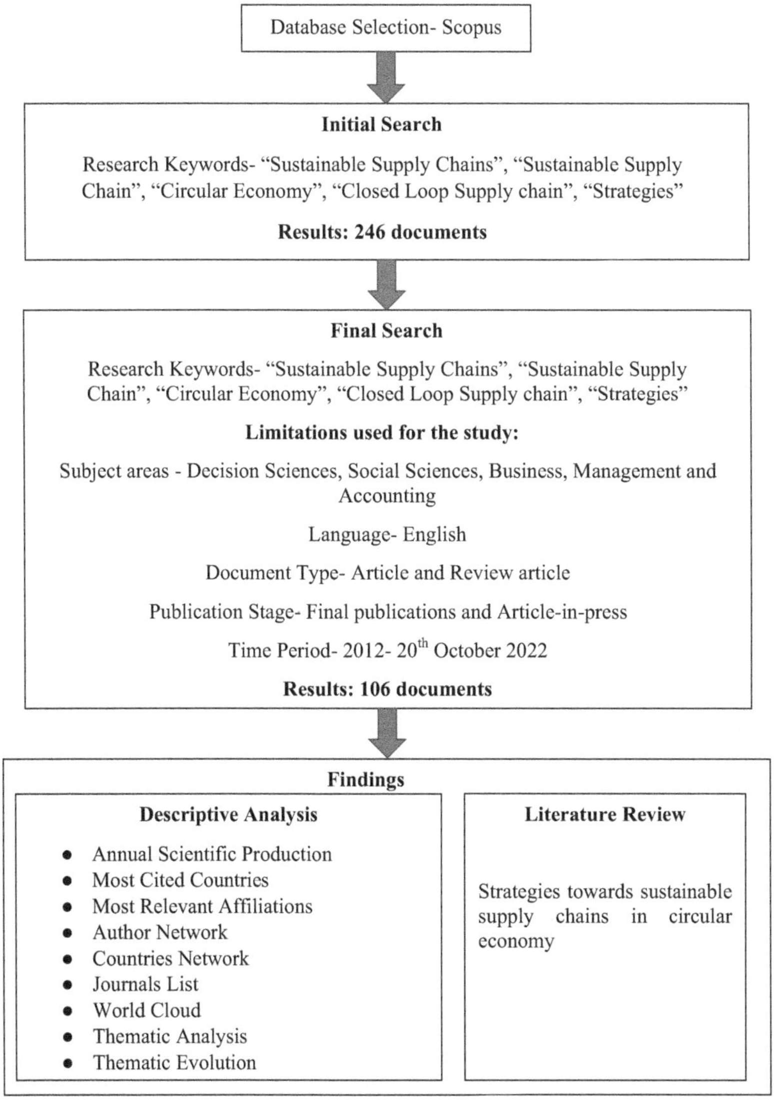

FIGURE 9.1 Flow chart of research methodology.

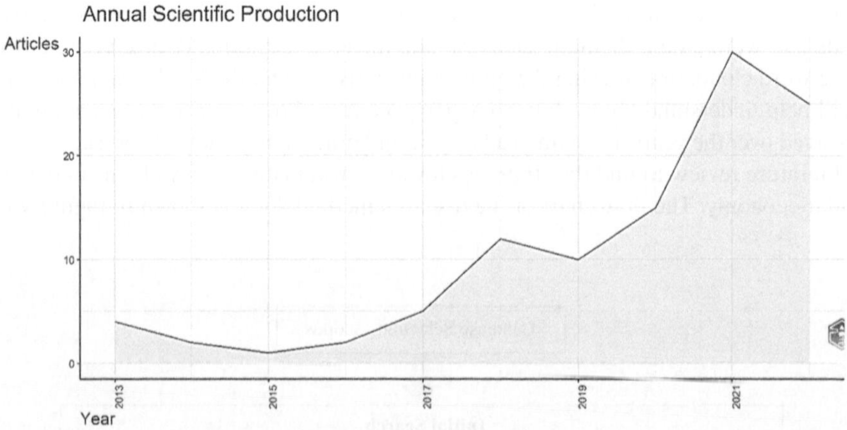

FIGURE 9.2 Annual scientific production (2012 to 20 October 2022). (Source: Bibliometric-based analysis using biblioshiny platform.)

9.3 RESULTS OF THE STUDY

9.3.1 Descriptive Analysis

9.3.1.1 Annual Scientific Production

This section provides the descriptive analysis of the study, including the annual scientific production, top ten most cited countries, top ten relevant affiliations, journals list, and the word cloud. The annual scientific production has been increasing over the years. The primary growth is shown after 2019, wherein in 2019 there were ten articles published on this topic. In 2020, 2021, and 2022 (up to 20 October 2022), 15, 30, and 25 papers were published, respectively. The annual scientific production is shown in Figure 9.2.

9.3.1.2 Most Cited Countries

The top ten most cited countries are shown in Figure 9.3. The study reveals that France is the most cited country.

9.3.1.3 Most Relevant Affiliations

The top ten relevant affiliations for the study are shown in Figure 9.4. Huazhong University of Science and Technology leads as the most relevant affiliation in our study.

9.3.1.4 Author Network

The author network is drawn using the VOSviewer software. The maximum number of authors per document is 25. The minimum number of documents of an author is 1. Out of 325 authors, 325 meet the threshold limit. Figure 9.5a shows the author network visualization. Figure 9.5b shows the author network overlay visualization.

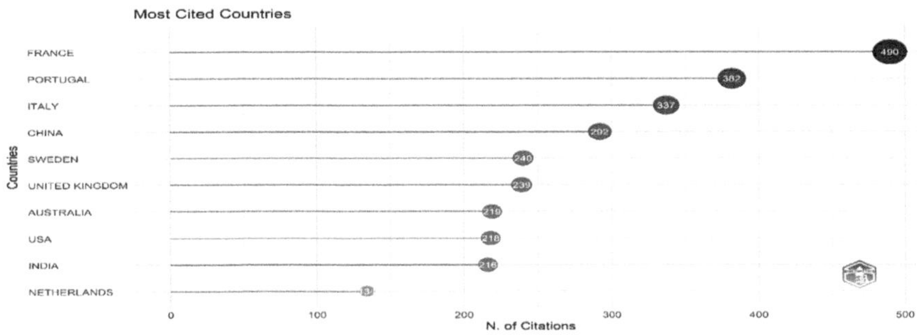

FIGURE 9.3 Most cited countries. (Source: Bibliometric-based analysis using biblioshiny platform.)

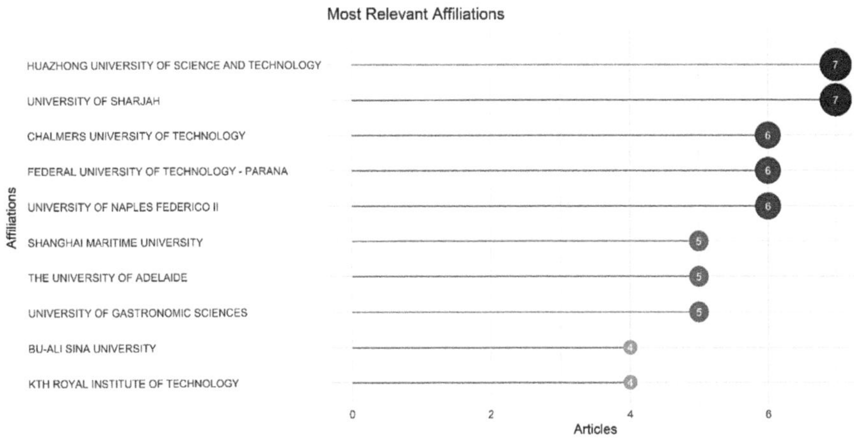

FIGURE 9.4 Most relevant affiliations. (Source: Bibliometric-based analysis using biblioshiny platform.)

9.3.1.5 Countries Network

The countries network is drawn using the VOSviewer software. The maximum number of authors per document is 25. The minimum number of documents in a country is 2. Out of 45 countries, 25 meet the threshold limit. Figure 9.6a shows the countries network visualization. Figure 9.6b shows the countries overlay visualization.

9.3.1.6 Journals List

Among the journals list (shown in Table 9.1), the *Journal of Cleaner Production* has published the highest number of articles in this area.

9.3.1.7 Word Cloud

The keywords reflect the main content of the research work. The keywords are usually represented as phrases or nouns. Figure 9.7a illustrates the word cloud drawn

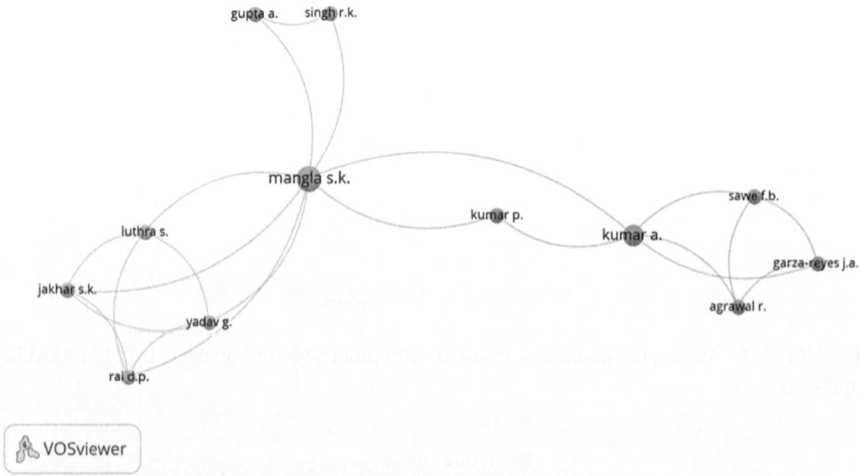

FIGURE 9.5A Author network (visualization). (Source: Bibliometric-based analysis using VOSviewer software.)

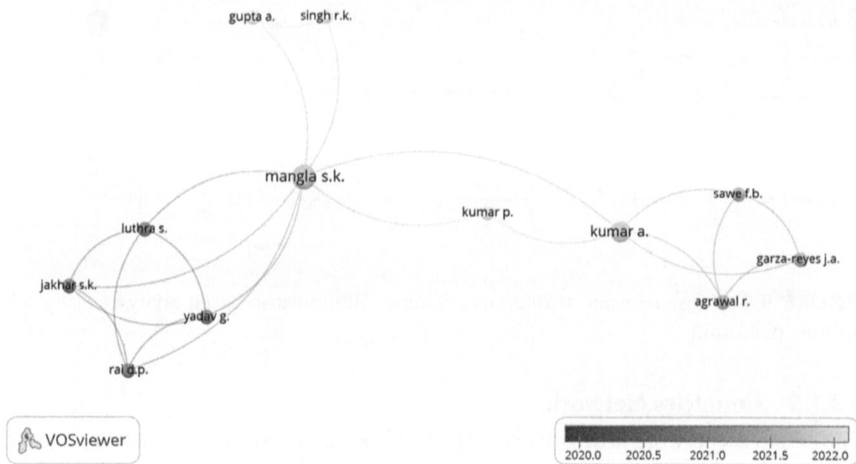

FIGURE 9.5B Author network (overlay visualization). (Source: Bibliometric-based analysis using VOSviewer software.)

from the biblioshiny webpage. Figures 9.7b and 9.7c are drawn from VOSviewer software. Figure 9.7b represents the word cloud network visualization, whereas Figure 9.7c highlights word cloud network overlay visualization. In Figure 9.7b, the study reveals that the top keyword used for the analysis is "supply chains", with a frequency of 55, followed by "sustainable development", with a frequency of 54. "Supply chain management", "circular economy", "closed-loop supply chain", and

FIGURE 9.6A Countries network (visualization). (Source: Bibliometric-based analysis using VOSviewer software.)

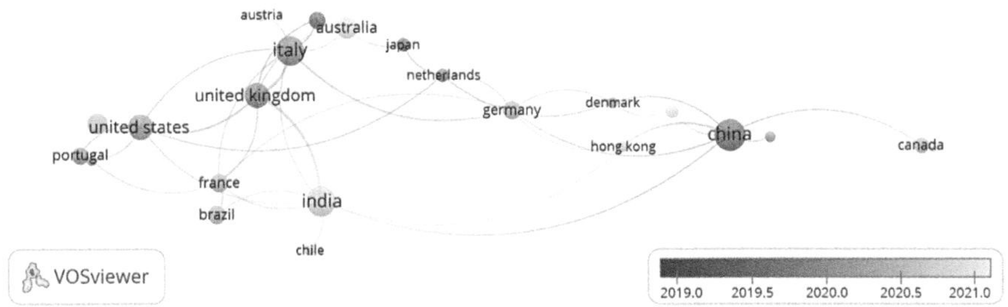

FIGURE 9.6B Countries network (overlay visualization). (Source: Bibliometric-based analysis using VOSviewer software.)

"recycling" had a frequency of 30, 28, 27, and 15, respectively. Further, "commerce" and "sustainability" follows a frequency of 13, whereas "decision making" and "environmental impact" follow a frequency of 12.

The keyword co-occurrence extracted from VOSviewer is drawn using the minimum number of occurrences of a keyword to be 5. Out of the 964 keywords, 40 keywords meet the threshold limit. For each of the keywords, the total strength of the co-occurrence links with other keywords are calculated. Then, the keywords with the greatest total strength are selected. Here, 40 keywords are chosen (Table 9.2).

9.3.1.8 Thematic Analysis

The thematic analysis is drawn using the biblioshiny webpage (Figure 9.8). The four quadrants represent different themes. The first quadrant, the upper right quadrant, is known as "motor themes" and is considered "underdeveloped". The second quadrant, the upper left quadrant, is known as "niche themes" and is considered marginally significant. The third quadrant, the lower left quadrant, is known as "declining or emerging themes" and is regarded as "weakly developed". Finally, the fourth quadrant, the lower right quadrant, is known as the "basic themes" and is considered "underdeveloped". The motor themes are the decision-making in the closed-loop

TABLE 9.1
Journal Sources for Articles Considered for the Study

Journal	Number of Articles
Journal of Cleaner Production	35
Sustainability (Switzerland)	10
International Journal of Production Research	6
Business Strategy and the Environment	5
International Journal of Production Economics	4
Environment, Development and Sustainability	3
IEEE Engineering Management Review	3
Journal of Enterprise Information Management	3
Journal of Industrial Ecology	3
Textile Outlook International	3
Annals of Operations Research	2
RAIRO - Operations Research	2
Resources Policy	2
Cleaner Logistics and Supply Chain	1
Computational and Mathematical Organization Theory	1
Economia Politica	1
Economic Annals-XXI	1
Emerald Emerging Markets Case Studies	1
Engineering Economics	1
Environmental Impact Assessment Review	1
Flexible Services and Manufacturing Journal	1
International Journal of Business Innovation and Research	1
International Journal of Logistics Systems and Management	1
International Journal of Productivity and Performance Management	1
International Journal of Supply Chain Management	1
Journal of Grey System	1
Journal of Industrial Engineering and Management	1
Journal of Japan Industrial Management Association	1
Journal of Manufacturing Technology Management	1
Journal of Retailing and Consumer Services	1
Journal of Water Sanitation and Hygiene for Development	1
Kybernetes	1
Management Decision	1
Omega (United Kingdom)	1
Production Planning and Control	1
Supply Chain Management	1
Sustainable Development	1
Transportation Research Part D: Transport and Environment	1

Source: Bibliometric-based analysis using biblioshiny platform.

FIGURE 9.7A Word cloud. (Source: Bibliometric-based analysis using biblioshiny platform.)

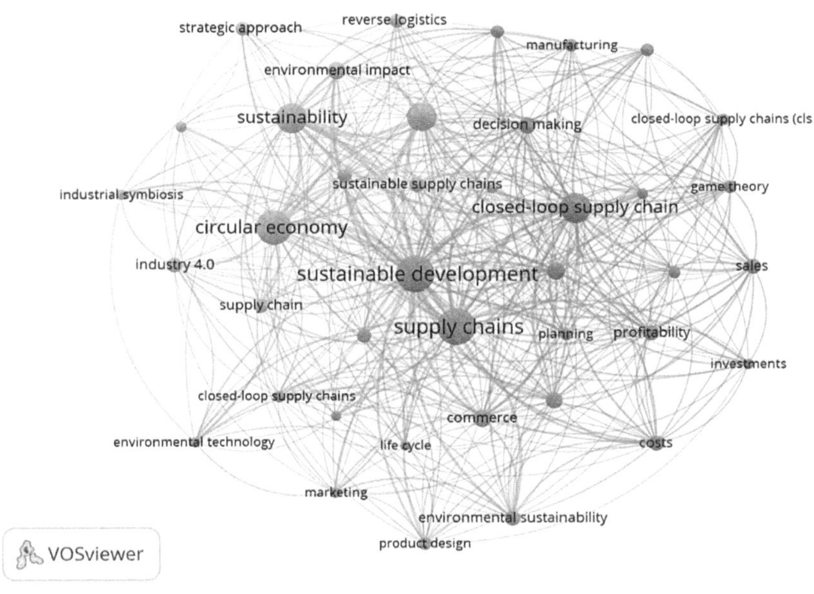

FIGURE 9.7B Word cloud network visualization. (Source: Bibliometric-based analysis using VOSviewer software.)

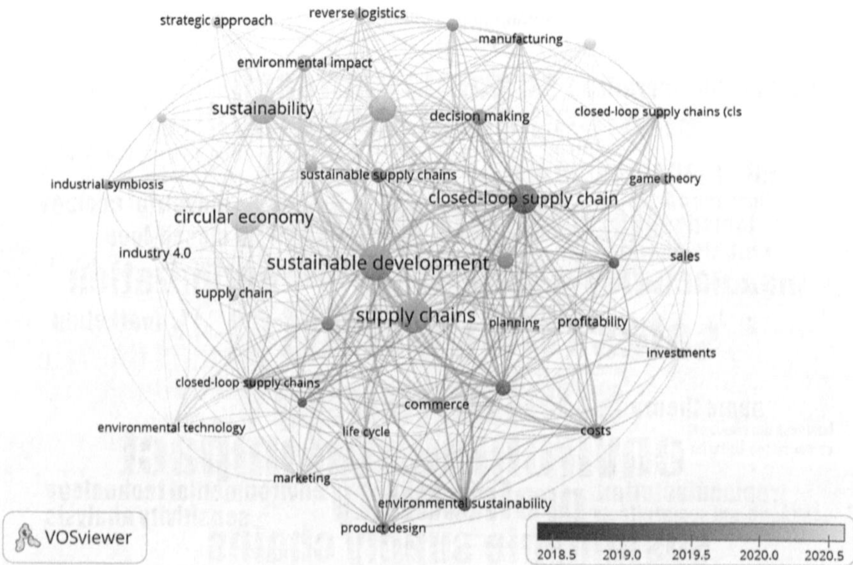

FIGURE 9.7C Word cloud network overlay visualization. (Source: Bibliometric-based analysis using VOSviewer software.)

supply chain theme. The pink colour highlights the sustainable development goals and circular economy theme. In the second quadrant, the niche theme is shown in orange and light blue. The orange colour highlights the circular economy in the food industry. The light blue colour highlights the past review works of literature on the circular economy. The third quadrant shows the declining or emerging themes using dark blue and brown colours. The dark blue colour highlights the quantitative analysis and the circular economy. The brown colour cluster highlights circular economy applications in the textile industry. The fourth quadrant reveals the basic themes highlighted with purple and light orange. The purple colour highlights the circular economy–based supply chains. The light orange highlights strategic management and circular economy.

9.3.1.9 Thematic Evolution

This section discusses the thematic evolution of the topic using the past ten years' data. The thematic evolution starts from the year 2013 to 20 October 2022. The number of cutting points in the time slicing is 2. Therefore, the cutting years selected for the study are 2017 and 2021. Figure 9.9a shows the thematic evolution. In 2017, the relevant keywords were logistics, supply chain management, closed-loop supply chain, and recycling. In 2018–2020, the relevant keywords were closed-loop supply chain, climatic change, logistics, optimization, sustainability, supply chain management, supply chains, numerical methods, and environmental management. In 2022, the relevant keywords were climatic change, supply chains, optimization, sustainability, and electronics industry.

TABLE 9.2
Keyword Occurrences and Total Link Strength

Keyword	Occurrences	Total Link Strength
Sustainable development	55	291
Supply chains	55	276
Closed-loop supply chain	36	200
Circular economy	47	187
Supply chain management	32	162
Sustainability	36	156
Commerce	13	95
Manufacture	11	79
Recycling	13	78
Profitability	10	66
Sales	10	66
Environmental impact	12	65
Costs	9	60
Environmental sustainability	9	58
Decision making	11	57
Sustainable supply chains	9	57
Remanufacturing	8	50
Game theory	7	47
Closed-loop supply chains	6	44
Waste management	8	43
Industry 4.0	10	42
Product design	7	41
Closed-loop supply chains (CLSC)	6	40
Supply chain	8	40
Closed-loop	5	39
Reverse logistics	8	37
Environmental management	6	36
Planning	6	36
Sensitivity analysis	7	35
Sustainable operations	6	35
Environmental technology	5	34
Marketing	5	34
Optimization	7	34
Manufacturing	7	33
Investments	5	32
Strategic approach	8	30
Industrial symbiosis	5	28
Life cycle	5	25
Sustainable manufacturing	5	24
Sustainable supply chain management	5	24

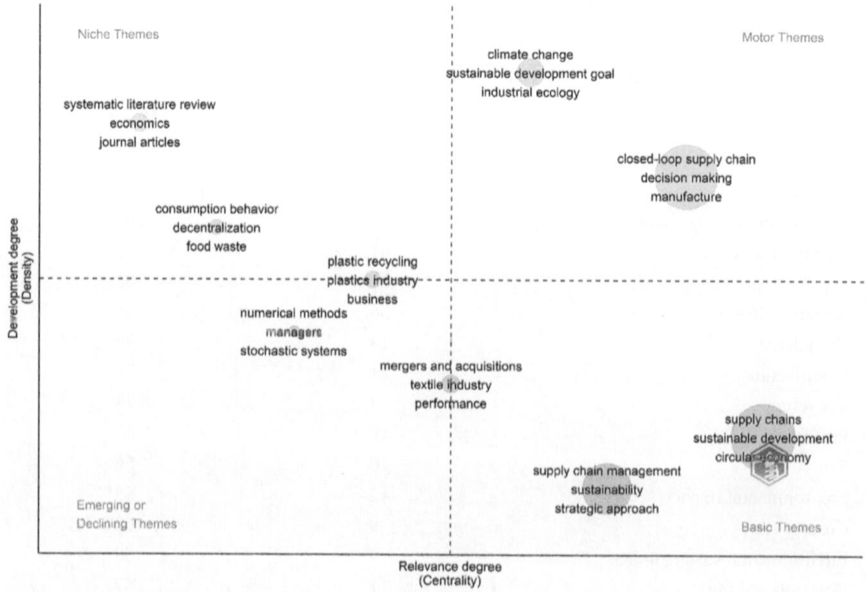

FIGURE 9.8 Thematic analysis. (Source: Bibliometric-based analysis using biblioshiny platform.)

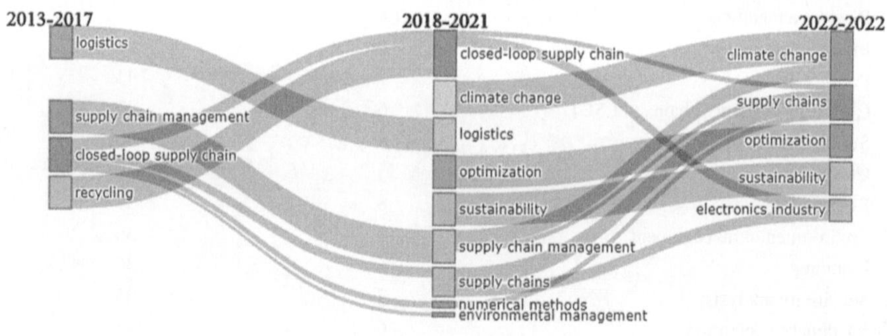

FIGURE 9.9A Thematic evolution. (Source: Bibliometric-based analysis using biblioshiny platform.)

Time slice 1 examines the topics of the past literature up to 2017. The significant themes under the motor theme indicates the sustainability development and closed-loop supply chain In the emerging or declining theme, recycling and logistics play an essential role in the circular economy context. Figure 9.9b shows the time slice 1 in the thematic evolution up to 2017.

Time slice 2 examines the topics of the past literature during 2018 to 2021. The motor themes are circular economy and environmental sustainability, and decision-making in the circular economy era. The niche themes are the food industry using

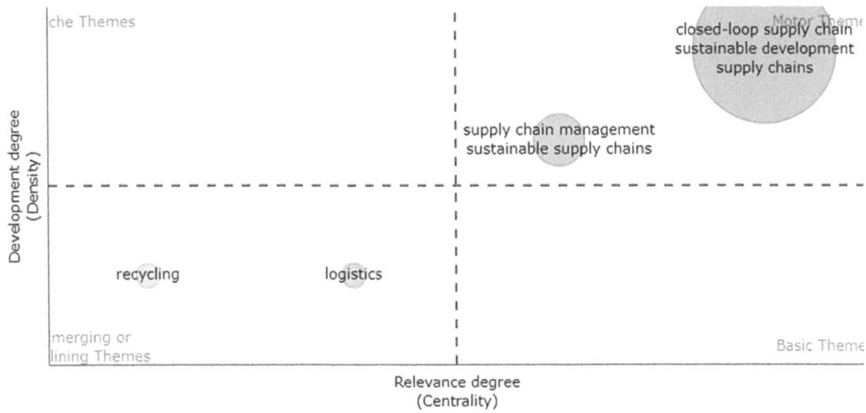

FIGURE 9.9B Thematic evolution (time slice 1). (Source: Bibliometric-based analysis using biblioshiny platform.)

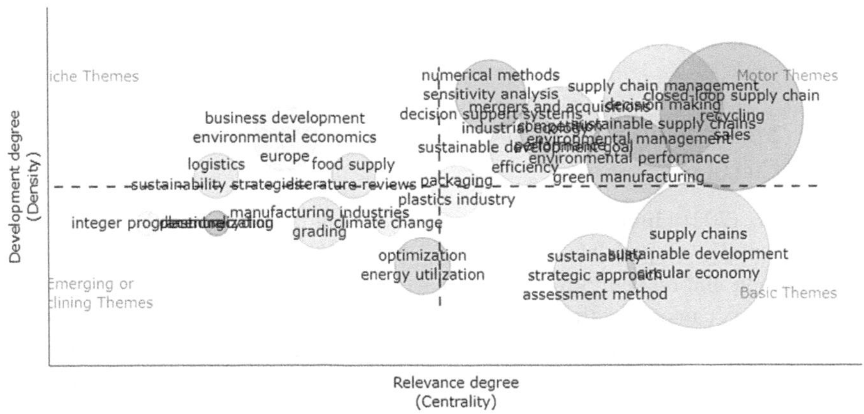

FIGURE 9.9C Thematic evolution (time slice 2). (Source: Bibliometric-based analysis using biblioshiny platform.)

circular economy and business developments in the circular economy era. The emerging or declining themes are energy management using closed-loop supply chains. The basic themes are circular economy–based supply chains. Figure 9.9c shows time slice 2 in the thematic evolution from 2018 to 2021.

Time slice 3 reveals the trending themes for the year 2022. This section plays a significant role in describing the topics which future researchers can take. The emerging or declining themes are optimization and collaboration in the circular economy. Regarding climatic changes, reverse logistics is considered a niche theme in the circular economy context. Future studies can work on how collaboration helps to attain closed-loop supply chains and how organizations should develop strategies

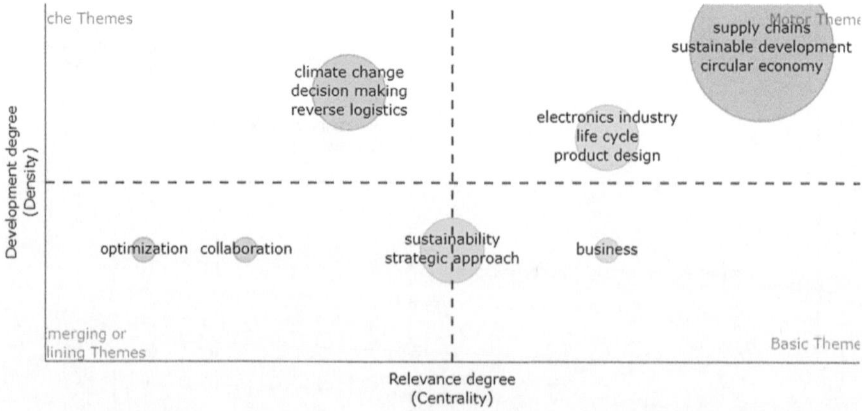

FIGURE 9.9D Thematic evolution (time slice 3). (Source: Bibliometric-based analysis using biblioshiny platform.)

to help firms implement a circular economy. Figure 9.9d shows the time slice 3 in the thematic evolution for 2022.

9.3.2 Strategies towards Sustainable Supply Chains in the Circular Economy

Feng et al. (2021) highlighted two strategies manufacturers could use to adopt environmentally responsible closed-loop supply chain models. One strategy is authorization and outsourcing for third-party manufacturers. Here, the third-party manufacturers can collect the end-of-life products and remanufacture them. A study by DHL (2022) highlighted that halving overproduction could reduce fashion waste by almost 10 million tons. Hence, on-demand production can be used to achieve circular economy principles. Assessing the product portfolio is another strategy which can be adopted (Capgemini, 2021). The products causing severe environmental harm can be eliminated or reduced from the production stages. For example, while assessing the sustainability reports and products portfolio, Puma, the world's leading sports brand, realized that the products made from cotton, leather, and rubber created a high negative environmental impact (Puma, 2011). As a result of the survey, the brand has started looking for alternative options for these raw materials to reduce greenhouse emissions and waste generation (Financial Times, 2012). To attain circular economy–based sustainable supply chains, firms must invest in building infrastructures to support recycling practices, identify new partners that can help implement circular economy, and enable transparency across partners (Capgemini, 2021). Other strategies include increasing the durability of the products, reducing waste materials, and increasing the cascading and reverse cycles (Elia et al., 2017). Figure 9.10 shows the strategies towards sustainable supply chains in the circular economy discussed in this chapter.

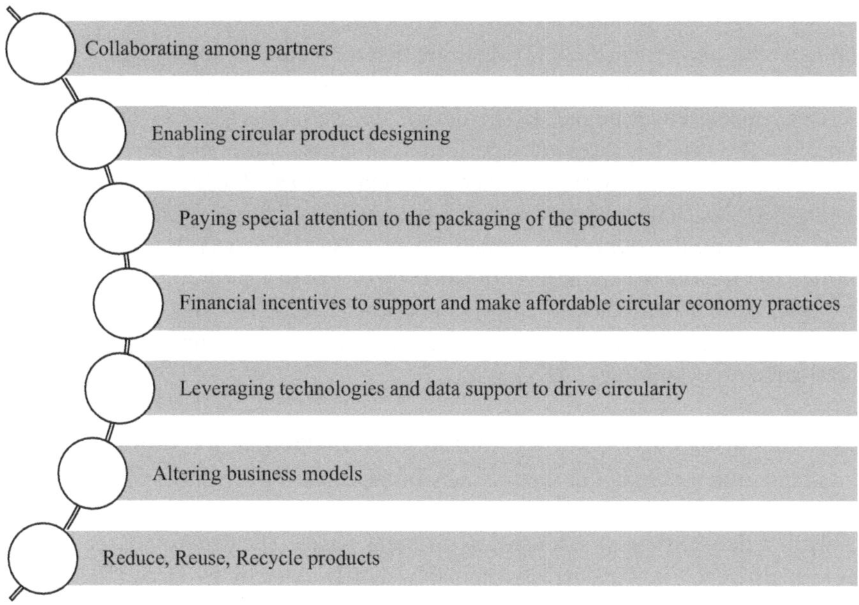

FIGURE 9.10 Strategies towards sustainable supply chains in the circular economy.

9.3.2.1 Collaborating among Partners

Pinho Santos and Proença (2022) revealed that collaboration among partners is essential to achieve sustainable supply chains. The collaborations are performed among different stages of the sustainable supply chain. The study highlights that sustainability should be integrated into all supply chain processes and at various supply chain stages. Hence, collaboration among partners is vital to attaining a sustainable supply chain. For example, IKEA collaborates with suppliers who follow sustainable practices and work for animal welfare (IKEA, 2022). Through more collaborations, organizations must try to close the loop wherever possible (Capgemini, 2021). Multiple sourcing and collaborating with suppliers are overall strategies to attain a circular economy. Multiple sourcing helps the firm achieve higher service levels than the single-sourcing method (Namdar et al., 2018). Collaboration can be established among the same industries. Regional, national, or global alliances can be made among organizations to achieve circular economy practices. These alliances will help the firm share information, resource, and knowledge. Collaborations can be made with start-ups, motivating them towards the circularity goals. Cross-industry collaborations can also be made. For example, in 2021, in the UK, a Flexible Plastic Fund initiative was made by collaborating with the top five consumer goods brands: Unilever, Nestlé, Mondelēz International, Mars, and PepsiCo. The collaboration was made with the motive to incentivize flexible plastics recycling. These practices include recycling pouches, plastic bags, and wrappers. This collaboration brings together the recyclers, consumers, retailers, local authorities, and manufacturers (Flexible Plastic Fund, 2022).

9.3.2.2 Enabling Circular Product Designing

Pinho Santos and Proença (2022) highlight that for a circular supply chain, rethinking product design from the initial stage is an important and essential step for manufacturing firms. For example, IKEA designs all its products using 100% circular concepts. The firm has targeted to perform responsible sourcing and use only recyclable and renewable materials for its production by 2030 (IKEA, 2022). In another instance, DSM–Niaga redesigned its carpet products from scratch (Niaga, 2017). Around "1.6 million tonnes of carpets go to waste every year" in the European Union. The materials used in traditional carpets are difficult to recover and reuse. DSM–Niaga has redesigned its carpets to be wholly recycled using Niaga's technology. The new design made the carpet lighter, improving the inner air quality and installation process (Niaga, 2017).

9.3.2.3 Paying Particular Attention to the Packaging of the Products

Looking at the packaging of the products in the direct-to-consumer market plays a vital role in attaining a circular economy. The firm must use packaging which can be easily recycled, reused, or recovered at different stages. The disposal of the packaging can efficiently be conducted sustainably shortly. A study by DHL (2022) highlighted that external packaging contributes to 30% of e-commerce emissions. For example, Apeel, a fruits and vegetable selling firm, has eliminated single-use plastic from packaging. Rather than using plastics, the firm uses a "peel" to package the products (Apeel, 2022). The peel comprises glycerolipids and lipids, made up of natural substances that save the fruits and vegetables from rotting. The peel maintains the moisture and keeps the oxygen out. It increases the product's shelf life and helps the firm deal with food waste (Apeel, 2022). This packaging positively contributes to maintaining the fruits' and vegetables' appearance, flavour, texture, and nutrition during the journey to retail outlets and supermarkets. By changing the product's packaging, the firm can keep the product fresh three times longer and promote better quality food products on the market (Apeel, 2022). To lower single-use plastic packaging, the UK retailers Asda and Tesco started providing refilling services (ASDA, 2022; BBC, 2021). Under this service, customers can buy new products in old reusable packaging. Once the product is consumed, the packages can be returned to the firm, and products resold in the old package. To encourage customers, the company has started providing a refundable deposit to customers for choosing the same packaging options. In the refilling zones, customers can reuse the containers for everyday groceries like pasta, pet food, cereals, and laundry detergents (ASDA, 2022).

9.3.2.4 Financial Incentives to Support and Make Affordable Circular Economy Practices

Changing linear models to circular ones incurs higher costs and therefore needs high investments or financial incentives (Esposito et al., 2020). There are few countries which have started incentivizing green practices. For example, the Netherlands, Sweden, and Austria have reduced the value-added tax percentage for repairs for various products (Right to Repair, 2020). Vienna, a city in Austria, has established "a system of repair vouchers". Under this voucher scheme, consumers can cover up to 50% of a product's repair cost (IEEE Spectrum, 2021; European Consumer Centre France, 2021). In 2022, France introduced a repair fund to subsidize repairs. Under this scheme, the company is expected to contribute towards the cost of repair of products. This scheme is a part of the "Extended Producer Responsibility" scheme (European Consumer Centre France, 2021).

9.3.2.5 Leveraging Technologies and Data Support to Drive Circularity

In the era of Industry 4.0, many technologies are accepted in the manufacturing and service industries. Organizations must leverage user data and technology to accelerate circular economy practices (Capgemini, 2021). Industry 4.0 technologies include artificial intelligence (Manikandan et al., 2022), the Internet of Things, cloud computing, digital twins, (Ali et al., 2022) blockchain, additive manufacturing, augmented reality, and virtual reality. These technologies help the firm to keep track of the materials and product flow within the supply chains. By keeping an eye on the materials and product flows, these technologies help build a circular mindset to implement circular economy practices (Capgemini, 2021). Digital platforms, websites, and apps can be used to connect consumers and producers to allow the sharing of products. Such approaches can lead to reducing waste. As discussed in (Juyal et. al. (2020), technologies such as wireless networks, machine learning, sensors, and robotics contribute to circulating practices for manufacturing products and processing the materials. Virtual reality and augmented reality can be used to try products before buying them. Such exercises help in avoiding waste creation. Digital twins, the Internet of Things, and blockchain technology drive the value chain optimization; increases repair, reuse, and refurbishments; develop the products as a service for many components and final products. Radio frequency identification provides complete information about the product lifecycle, which helps optimize product lifecycle management.

9.3.2.6 Altering the Business Models

Organizations are altering their business models in this strategy to attain a sustainable supply chain-based circular economy. While altering the business models, organizations should redefine the revenue channels. The revenue channels should not solely focus on product sales but rather embrace the circular economy practices across the firms (Capgemini, 2021). Under this model, one way can be attaining ownership with the firm aiming to build a product–service system. For example, Signify (formerly Philips Lighting) is a leader in lighting production, light fixtures, and selling electric lights worldwide. To adopt a circular economy, the company has altered the business model from selling the lighting equipment to offering it in a contract of either five years or more. It is known as the "lightning-as-a-service" business model. In this way, the company retains the ownership of the equipment with them and offers it on a contract basis, measuring some of the parameters. These parameters are energy savings, light level, light uptime, etc. With the contract about to expire, the consumer can either renew the contract or hand the lighting equipment back to Signify for recycling or reuse. By altering the business model, Signify is also incentivizing manufacturers to manufacture durable, high-quality, and modular equipment that can last longer and quickly circulate in the economy. These practices also promote responsible production and consumption, energy and cost savings, reduce carbon footprints and wastes, and make the system more efficient (Signify, 2022).

In another example, thredUP, a leading clothing platform for women's and kids' apparel, has created an easy, more straightforward, innovative way to buy and sell secondhand clothes. To adopt a circular economy, the company believes in reselling and buying old clothes rather than manufacturing new clothes. The model adopted by thredUP is known as the "resale-as-a-service" business model. The company has adopted this model for various brands and retailers. To resell the clothes, the company first collects old clothes from sellers, then processes these old clothes for various quality inspections, analysis, pricing, storage, and categorizing to resell the product. In this manner, the buyers can buy high-quality products at low prices from thredUP. By doing so, the company can keep the clothes in more prolonged use, which is the central strategy of the circular economy. In addition, thredUP can save 100 billion litres of water, 3.6 megatons of CO_2 emissions, and 200 million kilos of waste materials. If any clothes fail to pass the quality inspection, then thredUP sells them to textile recyclers or secondhand clothing stores. Moreover, thredUP offers a "Clean Out Kit" for those selling their clothes for free on this platform. This kit includes thredUP online credits, charitable donation receipts, cash, and select partner credits (thredUp, 2022).

In another instance, Kaer, an air-conditioning product and the service-based company selling air conditioner systems to industrial and commercial real estate around Asia, has altered its business model. Rather than selling the air conditioner systems, the company has adopted a "cooling-as-a-service" business model. The company realized the rising temperatures and high demand for air-conditioning systems. This increased demand creates more global warming and the need for more cooling systems. Therefore, the firm altered its business model and allowed building owners to buy "cooling" rather than cooling equipment. Kaer asks building owners about their requirements for air-conditioning and fulfils them by designing, installing, and operating the cooling service. The cooling service is charged based on the consumption units. Kaer uses the Internet of Things and artificial intelligence technologies to optimize building operations while designing modular air-conditioning units. By altering the model, Kaer is helping in saving energy and costs. It further helps reduce greenhouse emissions to the environment (Kaer, 2022).

9.3.2.7 Reduce, Reuse, and Recycle Products

According to World Bank reports, globally, 2.01 billion tonnes of solid waste are annually produced (The World Bank, 2022). It is expected to increase to 3.40 billion tonnes by 2050 (The World Bank, 2022). The UN Global E-Waste Monitor highlights that e-waste has grown by 21% in the last five years (2015–2022) (DHL, 2022). The natural resources present across the globe are in limited quantity. But the rate at which they are consumed is increasing daily. To deal with such a practical problem, reducing, reusing, and recycling products is very important. Reduce, reuse, recycle is also the basis of the circular economy. In addition, it can be used as a strategy to follow the circular economy principles. In a study by DHL, the reports highlighted that there would be 75 million tons of additional recycling materials if all countries achieved top recycling rates (DHL, 2022). Many firms have started using this strategy and are practising implementing it.

The pioneer of reusing and recycling materials is the automobile industry. In 2018, 87% of automobile materials and parts were recycled and reused in the European Union. Also, 93% of automobile materials and components from scrapped light vehicles and cars are recovered and reused. Renault, the world's leading automobile manufacturer, has developed a network of 330 salvage firms in France to recycle automobile parts and materials. Currently, the cars made by Renault in Europe contain 10%–20% recycled plastics and 36% recycled materials. Automobile firms have also increased use of electric vehicles. An electric vehicle's battery can be recycled and reused easily. Volkswagen has built a plant in Germany to recycle electric vehicle batteries. BMW also focuses on reusing electric vehicle batteries. Electric vehicle batteries are also reused to support local grid balancing and wind energy generation.

In another instance, CoreCentric Solutions is a Chicago-based product renewal and reverse logistics firm. The firm manages product repairs, returns, and remanufactures. The company aims to capture the value from the returned and defective products. The firm has developed an infrastructure model wherein it manages its parts collection system and collaborates with the world's largest consumer goods manufacturers and retailers. In contrast to the redistribution networks, which create deep-value secondary markets and new sales channels, repair programmes lessen the need for product replacements. As a result, the infrastructure model provided by CoreCentric enables items to have a longer useful life, which increases their utility and value. The company saves 400,00 products and 1,000,000 service parts annually from dumping into landfills. In addition, it saves approximately 500 megawatts of energy annually and reduces the warranty costs for manufacturers.

Further, it conserves 85% of the energy and materials required to create new products (CoreCentrics Solutions, 2022). In another example, since the fiscal year 2021, IKEA has phased out all non-rechargeable alkaline batteries and replaced them with new rechargeable batteries. IKEA takes this action to reduce waste and save money in the long run (IKEA, 2022). Further, in the same year, IKEA launched a new forest-positive plan, wherein the company aims to use certified and recycled woods to meet the sustainability requirements and support the livelihood of people depending on the forests. Moreover, this step further helps mitigate global climate change issues (IKEA, 2022).

9.4 CONCLUSION, LIMITATION, AND FUTURE SCOPE

A circular economy-enabled sustainable supply chain provides many benefits and opportunities to optimize finite resources, reduce negative climatic impacts, and improve responsible production and consumption patterns. There are a few challenges to adopting the circular economy, yet many firms are embracing the circular economy practices. Our study discusses the circular economy's strategies towards a sustainable supply chain. In the first half of our research, we conducted a bibliometric analysis using the biblioshiny webpage and the VOSviewer software. Here, we highlighted annual scientific production, most cited countries, most relevant affiliations, author network, countries network, journals list, word cloud, thematic analysis, and thematic evolution.

Further, we examined the strategies towards sustainable supply chains in the circular economy. These strategies include collaborating among partners, enabling circular product designing, paying particular attention to the packaging of products, financial incentives to support and make reasonable circular economy practices, leveraging technologies and data support to drive circularity, altering business models, and finally, reducing, reusing, and recycling the products. Our study provides multiple examples of firms adopting the strategies discussed in the text. The study findings will help academicians and practitioners understand the strategy in the context of a sustainable supply chain and circular economy.

Our study will help academicians and practitioners to understand the past ten years' trends in sustainable supply chains in a circular economy. Our study will further help young researchers better understand this area's emerging or declining basic, motor, and niche themes. In the future, studies can consider other databases such as the Web of Science, EBSCO, and Google Scholar for conducting similar studies. The strategies towards sustainable supply chains in the circular economy discussed in this work will help organizations implement these strategies and better understand alternate strategies.

REFERENCES

Ahmad, S., Mishra, S., & Sharma, V., 2023. "Green computing for sustainable future technologies and its applications". In Grima, S., Sood, K., & Özen, E. (Eds.), *Contemporary Studies of Risks in Emerging Technology, Part A (Emerald Studies in Finance, Insurance, and Risk Management)*, Emerald Publishing Limited, Bingley, pp. 241–256. https://doi.org/10.1108/978-1-80455-562-020231016.

Ajmani, P., Sharma, V., Samuel, P., Somasundaram, K., & Vidhya, V., 2022. "Patient behaviour analysis and social health predictions through IoMT". In *2022 10th International Conference on Reliability, Infocom Technologies and Optimization (Trends and Future Directions) (ICRITO)*, pp. 1–6. https://doi.org/10.1109/ICRITO56286.2022.9964846.

Ali, M. A., Balamurugan, B., Dhanaraj, R. K., & Sharma, V., 2022. "IoT and blockchain based smart agriculture monitoring and intelligence security system". In *2022 3rd International Conference on Computation, Automation and Knowledge Management (ICCAKM)*, Dubai, United Arab Emirates, pp. 1–7. https://doi.org/10.1109/ICCAKM54721.2022.9990243.

Ali, M. A., Balamurugan, B., & Sharma, V., 2022. "IoT and blockchain based intelligence security system for human detection using an improved ACO and heap algorithm". In *2022 2nd International Conference on Advance Computing and Innovative Technologies in Engineering (ICACITE)*, pp. 1792–1795. https://doi.org/10.1109/ICACITE53722.2022.9823827.

Apeel, 2022. "Working with nature to make food last longer: Apeel". Ellen Macarthur Foundation. Available at: https://ellenmacarthurfoundation.org/circular-examples/apeel. Last accessed on 20 October 2022.

ASDA, 2022. "ASDA to launch refill zones across metro stores". Available at: https://thisisretail.com.au/blog/asda-to-launch-refill-zones-across-more-stores/#:~:text=Asda

%20is%\20set%20to%20roll,laundry%20detergent%20and%20pet%20food. Last accessed on 20 October 2022.

Basheer, S., Singh, K. U., Sharma, V., Bhatia, S., Pande, N., & Kumar, A. (2023). "A robust NIfTI image authentication framework to ensure reliable and safe diagnosis." *PeerJ Computer Science*, 9, e1323.

BBC, 2021. "Tesco zero-waste trial launches at 10 stores in England". Available at: https://www.bbc.com/news/business-58549651. Last accessed on 20 October 2022.

Brydges, T., 2021. "Closing the loop on take, make, waste: Investigating circular economy practices in the Swedish fashion industry". *Journal of Cleaner Production*, 293, 126245.

Camacho-Otero, J., Boks, C., & Pettersen, I., 2018. "Consumption in the circular economy: A literature review". *Sustainability*, 10(8), 2758.

Capgemini, 2021. "Circular economy for a sustainable future: How organizations can empower customers and transitions to a circular economy". Available at: https://www.capgemini.com/wp-content/uploads/2021/11/Circular-Economy_11112021_v10_web-2-1.pdf. Last accessed on 20 October 2022.

CoreCentrics Solutions, 2022. "A second life – returns management, parts recovery and product repairs: CoreCentrics Solutions". Ellen Macarthur Foundation. Available at: A Second Life (ellenmacarthurfoundation.org). Last accessed on 20 October 2022.

DHL, 2022. "Delivering on circularity". Available at: https://www.dhl.com/global-en/home/insights-and-innovation/thought-leadership/white-papers/delivering-on-circularity.html#. Last accessed on 20 October 2022.

Durgalakshmi, K., Anbarasu, P., Karpagam, V., Venkatesh, A., Kannapiran, B., & Sharma, V., 2022. "Utilization of reduced switch components with different topologies in multi-level inverter for renewable energy applications-a detailed review". In *2022 5th International Conference on Contemporary Computing and Informatics (IC3I)*, Uttar Pradesh, India, pp. 913–920. https://doi.org/10.1109/IC3I56241.2022.10073430.

Elia, V., Gnoni, M. G., & Tornese, F., 2017. "Measuring circular economy strategies through index methods: A critical analysis". *Journal of Cleaner Production*, 142, 2741–2751.

EMF, 2013. "Towards the circular economy". Ellen Macarthur Foundation. Available at: https://www.ellenmacarthurfoundation.org/assets/downloads/publications/EllenMacArthur-Foundation-Towards-the-Circular-Economy-vol.1.pdf. Last accessed on 12 February 2020.

Esposito, B., Sessa, M. R., Sica, D., & Malandrino, O., 2020. "Towards circular economy in the agri-food sector. A systematic literature review". *Sustainability*, 12(18), 7401.

Esposito, M., Tse, T., & Soufani, K., 2018. "Introducing a circular economy: New thinking with new managerial and policy implications". *California Management Review*, 60(3), 5–19.

European Consumer Centre France, 2021. "Spare parts and repairs: An obligation in Europe?". Available at: https://www.europe-consommateurs.eu/en/shopping-internet/spare-parts-and-repairs.html#:~:text=The%20manufacturer%20has%20the%20obligation,(not%20necessarly%20for%20free). Last accessed on 20 October 2022.

Farooque, M., Zhang, A., Thürer, M., Qu, T., & Huisingh, D., 2019. "Circular supply chain management: A definition and structured literature review". *Journal of Cleaner Production*, 228, 882–900.

Fassio, F., Borda, I. E. P., Talpo, E., Savina, A., Rovera, F., Pieretto, O., & Zarri, D., 2022. "Assessing circular economy opportunities at the food supply chain level: The case of five piedmont product chains". *Sustainability*, 14(17), 10778.

Feng, Z., Xiao, T., & Robb, D. J., 2021. "Environmentally responsible closed-loop supply chain models with outsourcing and authorization options". *Journal of Cleaner Production*, 278, 123791.

Financial Times, 2012. "Puma to kick leather into touch". Available at: https://www.ft.com/content/579f4512-bc62-11e1-a836-00144feabdc0. Last accessed on 20 October 2022.

Flexible Plastic Fund, 2022. "Giving flexible plastic value". Available at: https://flexibleplasticfund.org.uk/about-the-fund. Last accessed on 20 October 2022.

Garfield, E., 1979. "Is citation analysis a legitimate evaluation tool?" *Scientometrics*, 1(4), 359–375.

Geissdoerfer, M., Savaget, P., Bocken, N. M., & Hultink, E. J., 2017. "The circular economy–A new sustainability paradigm?". *Journal of Cleaner Production*, 143, 757–768.

Goyal, S., Esposito, M., & Kapoor, A., 2016. "Circular economy business models in developing economies – Lessons from India on reduce, recycle, and reuse paradigms". *Thunderbird International Business Review*, 60(5), 729–740.

IEEE Spectrum, 2021. "Europe champions the right to repair are other regions next?" Available at: https://spectrum.ieee.org/right-to-repair-europe. Last accessed on 20 October 2022.

IKEA, 2022. "Sustainability – Caring for people and the planet". Available at: https://about.ikea.com/en/sustainability#:~:text=Responsible%20sourcing,human%20rights%20impact%2C%20among%20others. Last accessed on 20 October 2022.

Juyal, V., Saggar, R., & Pandey, N., 2020. "On exploiting dynamic trusted routing scheme in delay tolerant networks". *Wireless Personal Communications*, 112, 1705–1718.

Kaer, 2022. "Cooling as a service: Kaer". Ellen Macarthur Foundation. Available at: Cooling as a service: Kaer (ellenmacarthurfoundation.org). Last accessed on 20 October 2022.

Kalverkamp, M., & Young, S. B., 2019. "In support of open-loop supply chains: Expanding the scope of environmental sustainability in reverse supply chains". *Journal of Cleaner Production*, 214, 573–582.

Katiyar, R., Meena, P. L., Barua, M. K., Tibrewala, R., & Kumar, G., 2018. "Impact of sustainability and manufacturing practices on supply chain performance: Findings from an emerging economy". *International Journal of Production Economics*, 197, 303–316.

Korhonen, J., Honkasalo, A., & Seppälä, J., 2018. "Circular economy: The concept and its limitations". *Ecological Economics*, 143, 37–46.

Li, X., Huang, G., Chu, J., Zheng, B., & Huang, K., 2021. "How retailer co-opetition impacts pricing, collecting and coordination in a closed-loop supply chain". *Sustainability*, 13(18), 10025.

Lieder, M., & Rashid, A., 2016. "Towards circular economy implementation: A comprehensive review in context of manufacturing industry". *Journal of Cleaner Production*, 115, 36–51.

Manikandan, N., Ruby, D., Murali, S., & Sharma, V., 2022, October. "Performance analysis of DGA-driven botnets using artificial neural networks." In *2022 10th International Conference on Reliability, Infocom Technologies and Optimization (Trends and Future Directions) (ICRITO)*, pp. 1–6. IEEE.

Mardani, A., Kannan, D., Hooker, R. E., Ozkul, S., Alrasheedi, M., & Tirkolaee, E. B., 2020. "Evaluation of green and sustainable supply chain management using structural equation modelling: A systematic review of the state of the art literature and recommendations for future research". *Journal of Cleaner Production*, 249, 119383.

Maurya, A., & Sharma, V., 2022. "Facial emotion recognition using Keras and CNN". In *2022 2nd International Conference on Advance Computing and Innovative Technologies in Engineering (ICACITE)*, pp. 2539–2543. https://doi.org/10.1109/ICACITE53722.2022.9823480.

Mehrjerdi, Y. Z., & Shafiee, M., 2021. "A resilient and sustainable closed-loop supply chain using multiple sourcing and information sharing strategies". *Journal of Cleaner Production*, 289, 125141.

Merli, R., Preziosi, M., & Acampora, A., 2018. "How do scholars approach the circular economy? A systematic literature review". *Journal of Cleaner Production*, 178, 703e722.

Namdar, J., Li, X., Sawhney, R., & Pradhan, N., 2018. "Supply chain resilience for single and multiple sourcing in the presence of disruption risks". *International Journal of Production Research*, 56(6), 2339–2360.

Niaga, 2017. "Niaga®'s circular carpet manufacturing technology". Available at: http://www.circulary.eu/project/dsm-niaga/. Last accessed on 20 October 2022.

Pinho Santos, L., & Proença, J. F., 2022. "Developing return supply chain: A research on the automotive supply chain". *Sustainability*, 14(11), 6587.

Puma, 2011. "Puma completes first environmental profit and loss account which values impacts at €145 million". Available at: https://about.puma.com/en/newsroom/corporate-news/2011/11-16-11-first-environmental-profit-and-loss. Last accessed on 20 October 2022.

Right to Repair, 2020. "Austria makes repair more affordable". Available at: https://repair.eu/news/austria-launches-a-nation-wide-repair-bonus-scheme/#:~:text=Each%20voucher%20funds%2050%25%20of,a%20maximum%20of%20%E2%82%AC30. Last accessed on 20 October 2022.

Pushpalatha, N., Jabeera, S., Hemalatha, N., Sharma, V., Balusamy, B., & Yuvaraj, R., 2022. "A succinct summary of the solar MPPT utilizing a diverse optimizing compiler". In *2022 5th International Conference on Contemporary Computing and Informatics (IC3I)*, Uttar Pradesh, India, pp. 1177–1181. https://doi.org/10.1109/IC3I56241.2022.10072844.

Sassanelli, C., Rosa, P., Rocca, R., & Terzi, S., 2019. "Circular economy performance assessment methods: A systematic literature review". *Journal of Cleaner Production*, 229, 440e453.

Sharma, V., Balusamy, B., Thomas, J. J., & Atlas, L. G., eds, 2023. *Data Fabric Architectures: Web-Driven Applications*. Walter de Gruyter GmbH & Co KG.

Signify, 2022. "Why buy light bulbs when you can buy light? Signify". Ellen Macarthur Foundation. Available at: Why buy light bulbs when you can buy light? Signify (ellenmacarthurfoundation.org). Last accessed on 20 October 2022.

Singh, A., Dhanaraj, R. K., Ali, M. A., Balusamy, B., & Sharma, V., 2022. "Blockchain technology in biometric database system". In *2022 3rd International Conference on Computation, Automation and Knowledge Management (ICCAKM)*, Dubai, United Arab Emirates, pp. 1–6. https://doi.org/10.1109/ICCAKM54721.2022.9990133.

Soleimani, H., 2021. "A new sustainable closed-loop supply chain model for mining industry considering fixed-charged transportation: A case study in a travertine quarry". *Resources Policy*, 74, 101230.

Son, D., Kim, S., Park, H., & Jeong, B., 2018. "Closed-loop supply chain planning model of rare metals". *Sustainability*, 10, 1061.

Stewart, R., & Niero, M., 2018. "Circular economy in corporate sustainability strategies: A review of corporate sustainability reports in the fast-moving consumer goods sector". *Business Strategy and the Environment*, 27(7), 1005–1022.

The World Bank, 2022. "What a waste 2.0: A global snapshot of solid waste management to 2050". Available at: https://datatopics.worldbank.org/what-a-waste/trends_in_solid_waste_management.html. Last accessed on 20 October 2022.

thredUp, 2022. "Keeping clothes in use to reduce waste: thredUP". Ellen Macarthur Foundation. Available at: Keeping clothing in use to reduce waste: thredUP (ellenmacarthurfoundation.org). Last accessed on 20 October 2022.

UNCTAD, 2022. "Circular economy". United Nations Conference on Trade and Development. Available at: https://unctad.org/topic/trade-and-environment/circular-economy#:~

:text=A%20circular%20economy%20entails%20markets,economy%20or%20used%20more%20efficiently. Last accessed on 20 October 2022.

World Resources Institute, 2021. "5 opportunities of a circular economy". Available at: https://www.wri.org/insights/5-opportunities-circular-economy. Last accessed on 20 October 2022.

Yoo, S. H., & Cheong, T., 2021. "Inventory model for sustainable operations of a closed-loop supply chain: Role of a third-party refurbisher". *Journal of Cleaner Production*, 315, 127810.

Zhang, Z., Liu, S., & Niu, B., 2020. "Coordination mechanism of dual-channel closed-loop supply chains considering product quality and return". *Journal of Cleaner Production*, 248, 119273.

10 Research Potentials and Future Trends of Digital Sustainability

Shilpy Verma, Vandana Sharma, Vinay Kukreja, Achyut Shankar, and Ahmed A. Elngar

CONTENTS

10.1 Introduction 187
10.2 Sustainable Digital Technology 189
10.2.1 Trends in Digital Technologies: The Changing Landscape of the Future 190
10.3 How Digital Technology and Innovations Can Protect the World 191
10.3.1 Harnessing the Power of Data and Digital Technologies 193
10.4 Digital Expertise: A Way towards Sustainability in Business 194
10.4.1 Use Cases across Relevant Action Areas 194
10.4.1.1 Climate Change Challenges 194
10.4.1.2 Biodiversity and Conservation 195
10.4.1.3 Harnessing Potential of Technology to Improve the Health of the Ocean 195
10.4.1.4 Digital Technologies for Water Security, River Quality, and Drought Prediction 196
10.4.1.5 Digital Technologies for Pollution Reduction and Air Quality Alerts 197
10.4.1.6 Weather and Disaster Resilience Action Areas 197
10.5. Scope for Future Research 198
10.6 Conclusion 200
References 201

10.1 INTRODUCTION

The global crisis fuelled by systematic shocks caused by the coronavirus (COVID-19) pandemic has quickened the rate of digital technologies adoption. According to the United Nations Department of Economic and Social Affairs (UN DESA), digital solutions have played a crucial role, not only in expanding our understanding of virus biology and scientific research, but also enabling continued economic activities and provision of basic social services, including healthcare and education. Many of our

DOI: 10.1201/9781003348313-10

economies have avoided a complete shutdown in recent months as billions of people are asked to study or work from home thanks to digital technology. Not surprisingly, some countries have seen internet traffic surge by as much as 60% since the crisis began. Given the unprecedented extent to which our world relies on digital tools for prosperity and connectivity, digital technology should not merely be seen as a means of boosting economic performance, but its impact on the environment must also be assessed. Sustainable digitization can unlock the full potential of technology and addresses concerns over sustainability. In the past, all technology revolutions and transformations have brought some kind of prosperity and progress, and contributed to overall economic growth and development of the economy. However, several unprecedented consequences in terms of global warning, climate change, and environmental pollution have followed such transformations and revolutions. The current digital technology transformation is no exception to such consequences. Hence, digital technologies pose both opportunities and challenges for environmental sustainability. The implementation of digital technologies requires full comprehensive analysis of the economic benefits accruing from such implementation on one hand and environmental implications on the other hand. If environmental effects are left unchecked, overall losses to society from such implementation might exceed the benefits accruing from it. Bringing digital technology and environmentally sustainable practices together should be the forefront of strategic thinking for any economy.

By appropriately harnessing digital technologies, the digital transformation can be driven towards environmental sustainability to mitigate climate change–related risks and advance global sustainability. As we all are aware, information and data is crucial to monitor, protect, and spread awareness about overall planetary health, to inform markets, to influence consumer behaviour, and to reform governance systems. For example, there are multiple digital technologies such as artificial intelligence (AI), machine learning, and blockchain to prevent food wastage. The use of digital technologies helps reduce, prevent, and recycle food waste. These include technologies related to food recycling, traceability, preservation, packaging, storage, and food donations. In Bogota, the EatCloud platform, by applying digitization and AI, identifies the moments and places where food is being wasted along the entire supply chain. It digitally connects food manufacturers and retailers with food banks. Using AI and the real-time data provided by food suppliers, the platform identifies the ideal beneficiary for any kind of wasted food.

The wide range of water security challenges will require digital technology-based solutions, ranging from blockchain technologies to artificial intelligence, using mobile and internet applications to big data and remote sensing. The gap between supply and demand of water resources can be bridged with block chain technology. In India, state-wise analysis has shown that using block chain technology, issue of water quality, water loss, ground water restoration, and waste water treatment can be resolved [1].

According to the World Economic Forum [2], digital technology has the potential to cut greenhouse gas (GHG) emissions by 15% and transform the infrastructure sectors that contribute significantly to GHG emissions, such as the transport and power sectors. Industry 4.0 technologies such as big data can can ensure superior energy efficiency and optimization of operations in terms of high saturation of manufacturing equipment, lowwaste and scrap rate [3]. By applying a combination of advancements in digital crop, microbial genetics, agriculture, and electrification, it is possible to reduce GHG emissions by 71% (1744 kg CO2e/ha) from row crop agriculture in the next 15 years [4]. Based on sensor output, digital plant and soil health maps will be updated by real-time automated systems. By using robotics, a farmer can achieve positive environmental benefits by reducing energy consumption and waste production [5]. Moreover, innovations such as the Industrial Internet of Things, smart ecosystem, customized standardization, and asset automation have the potential to be optimized across many industries including the oil and gas industry. These innovations can prevent the loss of US$34 billion annually emanating from leaked methane, averting oil spills corresponding to approximately 230,000 barrels, saving about 800 million gallons of water, and reducing CO_2 emissions by approximately 1.3 billion tons [6].

The digital technology deployment can reduce carbon emissions by promoting new carbon reduction methods, eliminating or substituting the use of emissions-intensive technology [7]. Digital technologies such as artificial intelligence, cloud computing, and big data could have a "spillover effect" via a platform economy, shared networking, and business collaboration by neighbouring regions [8]. The spillover effect implies that digital technology adoption can promote environmental awareness and green technologies production. As a result, the digital technology in one region could enable carbon abatement in other regions. Digital technology adoption could enhance the public awareness of environmental protection by accelerating the transmission of environment-related information in surrounding regions.

10.2 SUSTAINABLE DIGITAL TECHNOLOGY

Digital technologies such as the Internet of Things (IoT) [9], blockchain [10], cloud computing [11], and AI are used to optimize the use of resources and has the potential of making distribution of scarce resources (such as land, labour, electricity, water) more efficient. It is also used to reduce GHG emissions, track and measure sustainability progress, promote innovation, and enable climate change solutions. We will analyze the potential and challenges of digitalization in achieving sustainability. Using real-life cases, we will investigate how digital technologies can be both an obstacle and enabler to global sustainability, which will enable devising appropriate digitalization strategies geared towards the achievement of sustainability.

Digital technology sustainability means ensuring that digital systems are managed, produced, designed, and used in a way that contributes towards sustainable development, improves climate resilience, and minimizes social and environmental impact. The global supply networks running on enterprise systems, cryptocurrency mining, and large data centres are some of the main contributors to global warming, placing the global carbon footprint of information and communication technologies on par with the aviation industry. This has led to increased scrutiny of digital supply chains for their use of rare earth minerals, conflict elements, and modern slavery, while e-waste has become the world's fastest growing waste stream.

Digital technology has the ability to advance all 17 of the UN Sustainable Development Goals – involving more than 50% of the total 169 subtargets, according to the Global e-Sustainability Initiative (GeSI). The ability of IoT and digital devices to transmit, capture, and generate data can have implications for sustainability. The increase in data quality and quantity with the help of advanced analytics provides insights to the efficiency of organization functioning. This helps identify energy consumption patterns and opportunities to reduce environmental impacts. Given this background, this study assesses the environmental impact of digital technologies, how the innovation can play a role in minimizing environmental damages arising from it, and the challenges associated with its implementation.

10.2.1 Trends in Digital Technologies: The Changing Landscape of the Future

Digital sustainability is the practice of using digital technologies to minimize the negative environmental impact associated with them. This includes reducing GHG emissions, energy consumption, carbon emissions, water use, and waste from within the business and its supply chains. According to a recent Capgemini Research Institute report [12], it is estimated that by 2025, the number of connected devices will reach 55.7 billion. This growth is driven by the falling cost of sensors and networking technology, as well as the increasing demand for connected devices. The benefits of IoT are vast and include everything from improved efficiency and productivity to greater oversight and control.

However, for all the benefits, there are also some downsides. IoT devices require a lot of energy to run. They need to be constantly connected to the internet and often use battery power, of which batteries can damage the environment if not disposed of properly. The number of devices in use also reduces a company's energy efficiency. The way companies conduct business has changed with the rise of remote work and international clientele. Businesses have had to adapt the way they operate, which has put a strain on the planet's resources. Although a study in [13, 14] suggests the implementation of green resources and green computing in sustainable development.

The global supply networks backed by large data centres, cryptocurrency mining, data-enabled analytics, and applications are some of the major contributors to global emissions. During the period from 2013 to 2018, the carbon emissions of digital technology has increased from 2.5% to 3.7%, which is much bigger than that the carbon footprint of the aviation industry (2.5%). A significant rise in data centre energy usage was observed due to greater adoption of cloud technology, which represents about 2% of global electricity today and is expected to increase to 8% by 2030. These trends are expected to get exacerbated by adoption of social media and digital services across the world. As we become more connected, the demand for the development and greater access to IoT has increased. During the period from 2016 to 2030, data centre energy consumption [15] will rise from 200 TWh to 2967 TWh [16]. The internet data centres market is expected to reach US$143.4 billion by 2027 from US$59.3 billion in 2020, increasing at a compound annual growth rate of 13.4% [17]. The internet causes indirect impact on pollution via its impact on electricity consumption. In order to mitigate the risks of the explosive growth in data consumption and devices, we need to give greater emphasis on sustainable solutions.

As outlined earlier, digital technologies can have a negative environmental consequence [18, 19]. By harnessing the advances in technology appropriately, digital technologies can offer groundbreaking opportunities to protect and monitor the environment in the most appropriate way. Emerging digital technologies such as AI, IoT, cloud computing, big data analytics, and mobile technologies can contribute to environmental sustainability. These technologies are expected to improve energy efficiency and enhance our ability to minimize pollution through control and surveillance of complicated processes. For example, by implementing sustainable business practices, digital technologies are expected to reduce carbon emissions and waste generation [20]. To identify waterborne diseases in waters, AI-based technology and smart water management systems can be used [21]. Data collected from such technology can develop and implement solutions and reduce water pollution. In addition, big data analytics have increased the capability to measure and trace the influence of various human activities on the environment [22]. Given the sustainability-related advantages associated with implementation of digital technologies, policymakers, governments, and businesses can enhance sustainability by incorporating digital technologies into various systems, processes, and activities.

Digital technology is expected to reduce carbon emissions worldwide by generating energy-saving benefits, developing transportation systems, smarter cities, next-generation distributed networks, and industrial processes. Emerging technologies such as robotics and automation, drones, blockchain, and AI contribute to the UN's Sustainable Development Goals by precisely monitoring and controlling complex processes, improving the ability to reduce pollution, and enhancing energy efficiency. Because real-time monitoring transparency in supply chains will improve, the environment can be more easily investigated and protected.

10.3 HOW DIGITAL TECHNOLOGY AND INNOVATIONS CAN PROTECT THE WORLD

According to the World Economic Forum (2022), digital technologies can abate GHG emissions in the top three contributing sectors – materials, mobility, and energy – by up to 20% by 2050 [23]. In the year 2020, the materials, energy, and mobility sectors contributed around 26%, 43%, and 24%, respectively, of the total GHG emissions (Figure 10.1). These sectors can utilize four enabling digital technologies to cut the carbon emissions of their operations and technologies underlying the foundation of their value chains: big data analytics; enabling technologies such as augmented reality, blockchain, cloud computing, and 5G; decision-making technologies such as machine learning/artificial intelligence; and control and sensing technologies such as automation [24], drones, and IoT.

Digital solutions can reduce greenhouse gas emissions by 7% in the materials sector by improving upstream production and mining, and enhancing efficiency and circularity of materials. The materials sector includes industries engaged in metal refining and mining, chemicals, construction, and forestry products. In the year

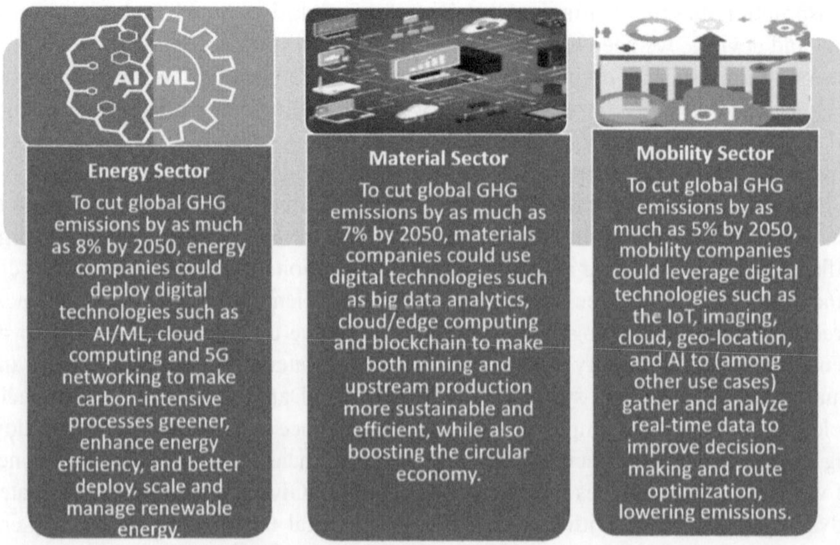

FIGURE 10.1 Decarbonizing the energy, materials, and mobility sectors with digital technologies.

2020, 21% of global emissions were the resultants of the creation of concrete, chemicals, plastics, and metals. The integration of digital technologies into the entire production stream can optimize existing processes, improve the efficiency of resource utilization, and reduce the wastage of resources.

The energy sector, which contributed 34% of global emissions in 2020, represents the largest decarbonization opportunities. In the energy sector, digital solutions can reduce emissions by up to 8% by managing and deploying renewable energy, improving the energy efficiency of buildings, and streamlining carbon-intensive processes.

The mobility sector, which includes the mobility of goods and people via rail, water, and air transport, contributed 19% of global emissions in 2020. A reduction in emissions from the sector requires a transition from fossil fuels–based energy consumption to green energy such as hydrogen. By increasing supply chain efficiency and streamlining travel routes, it is possible to switch from the burning of fossil fuels to the use of green molecules and cut emissions from the mobility sector by 5% by the year 2050.

Precision agriculture [25] is the notion where decision support technologies, robotics, automation, and IoT are deployed for making farming more sustainable, profitable, and efficient. For real-time data collection, precision agriculture uses tractor GPS, satellite imaging, and weather stations. This has resulted in water use reduction by 50–180 billion cubic metres, increased production by up to 300 million tonnes, and reduced production costs by US$40 billion–US$100 billion [26]. Application of AI in the agriculture sector would ensure on a timely basis informed decision-making, automated data collection, and corrective actions. The application of digital technologies ensures optimal agriculture inputs use, lower use of pesticides and fertilizers, detection of crop diseases, and conservation of natural resources such as land, water, and soil [27].

10.3.1 Harnessing the Power of Data and Digital Technologies

As per the United Nations Environment Programme (UNEP), in the years to come, a digital technology and innovation solution can help the world understand environmental hazards and accelerate global environmental action. A study by UNEP has shown that there is not enough data available for measurement or assessment of the progress of 68% of the sustainable development goals indicators. To track air quality at the local, national, and global levels, the Global Environment Monitoring System for Air (GEMS Air) is the tool used by UNEP for real-time monitoring. GEMS Air, covering about 5000 cities worldwide, is the largest air pollution network. In 2020, the platform had more than 50 million users and its data are transmitted to digital billboards to warn people about air quality risks in real time. This information is used to estimate local air pollution and age-specific exposure to air pollution in each country. Building on lessons from GEMS Air, UNEP aims to implement this facility into mobile phone health applications. In order to harness the power of data and digital technologies such as cloud computing, AI, and earth observation, UNEP has built three lighthouse digital platforms to monitor water bodies, methane emissions, and natural systems [28].

The Freshwater Ecosystem Explorer is a high-resolution geospatial data platform that depicts the state of rivers and lakes for all countries across the globe and helps understand the extent to which dynamics of freshwater ecosystems has changed over time. This platform is used to track/assess freshwater quality and intended to drive efforts to restore freshwater ecosystems. Data on the quality of seasonal and permanent surface water bodies, aquifers, wetlands, and mangroves can be downloaded at national, regional level, and watershed scales, and visualized using geospatial maps and associated data graphics.

Methane is a greenhouse gas contributing about 25% towards global warming. In order to catalyze the methane emission mitigation actions and create an independent and sound scientific basis for methane emission reduction, UNEP launched the International Methane Emissions Observatory (IMEO) in the year 2022 at G20 Summit. The observatory is designed to gather data from various sources including ground-based sensors, satellites, and scientific studies, and to integrate diverse methane emissions data to identify the origins of methane emissions. By providing reliable and real-time data on the quantity and locations of methane emissions, the observatory can scale up the concerted efforts needed to combat climate change.

The United Nations Biodiversity Lab 2.0 under the ambit of UNEP is a coalition of technological innovators and data providers intended to create the digital ecosystem for natural systems. The platform features spatial data and state-of-the art maps providing information on the state of biodiversity, the effects of climate change, and the scale of human development that helps policymakers generate insights for sustainable development and nature. The policy decisions backed by spatial data can facilitate the implementation of nature-based solutions.

10.4 DIGITAL EXPERTISE: A WAY TOWARDS SUSTAINABILITY IN BUSINESS

Combining sustainable practices with digital competence should form a crucial aspect of strategic thinking of any business. It is hard for any business to reduce its carbon footprint or dispose waste without having any understanding of sustainability. Emerging technologies can trigger innovation, which in turn can increase productivity, and help to identify the waste of resources across the supply chain [29].

The World Economic Forum, in collaboration with PwC, published a report in 2018 that identified 30 ways in which AI technologies and innovations could drive transformation across the natural resource and environment system security agenda, and contribute towards sustainability [30]. By combining big data, cloud computing, and AI capabilities, natural disaster resilience planning, climate modelling, and energy use forecasting can be improved. Next, we will discuss how digital technologies can address the most pressing environmental challenges.

Blockchain technology, big data analytics, remote sensing, and IoT sensors, all support the monitoring of energy and waste flows [31]. AI algorithms, cloud computing, and big data analytics can be used to predict natural disasters such as floods, droughts, forest fires, and earthquakes, and preserve and monitor endangered species on land. Smart grids driven by AI can reduce GHG emissions by monitoring trends in energy consumption. Satellite imagery and data collection tools can monitor the pollution and health of ocean habitats and marine ecosystems, prevent overfishing, thus ensure ocean sustainability. IoT, combined with smart surveillance and security systems, can be used to monitor animal behaviours and prevent poaching attempts by tracing endangered animals. Blockchain technology has numerous applications in the sustainability of the environment, according to the report "Building Block(chain) s for a Better Planet" [32]. For example, blockchain technology through transparent and traceable supply chains can prevent overfishing and impact positively marine biodiversity. It can also contribute towards sustainable resource management, including energy consumption and water. Digital technologies have the immense potential to unlock the solutions or way in which environmental problems are tackled.

The next section discusses use cases across relevant action areas pertaining to climate change, air pollution, biodiversity, ocean species, water pollution, and weather and disaster resilience to demonstrate how each area stands to benefit from the application of digital technologies and how research in these areas can help in the attainment of sustainability goals.

10.4.1 Use Cases across Relevant Action Areas

10.4.1.1 Climate Change Challenges

Digital technologies can play a crucial role in addressing the challenges associated with climate change and contribute significantly to improving sustainability. We might develop applications and technologies that could reduce energy consumption and emissions. Organizations can use AI-based algorithms-driven systems to plan and implement sustainable projects, track sustainability efforts, and optimize performance.

There are many action areas pertaining to climate change that can benefit from AI. For example, in case of smart cities and homes, machine learning algorithms can be used to examine data of millions of meters and smart sensors to make predictions on energy usage cost and requirements within buildings. In clean power, increasing efficiency and decreasing unpredictability, realizing the full potential of "smart grids", storage and use of renewable energy, and power balancing, machine learning is being used to map demand in real-time and energy generation. Light sensors can be used to switch on and off or decrease lighting levels automatically outdoors and indoors, allowing electric light levels to automatically function in response to changes in ambient light. In this way the energy consumption for indoor lighting can be reduced. AI-guided autonomous vehicles (AVs) present substantial opportunities for urban transport GHG reductions. Examples include autonomous ride-sharing services, eco-driving algorithms that prioritize energy efficiency, and route optimization that reduces driving congestion and miles. Digital technologies could contribute to climate change mitigation and resources conservation through energy-efficient processes [33]. Hence, researchers can guide organizations and individuals in analyzing and tracking their carbon footprint, and developing strategies and plans to achieve sustainability.

10.4.1.2 Biodiversity and Conservation

There are numerous ways in which we can conserve and monitor our habitats using AI. For example, combined with satellite imagery, applications of AI can identify forest and vegetation degradation, land-use changes, and monitor floods. AI-based applications offer a considerable scope to facilitate the precise management of forest habitats, monitoring of landscapes, and mitigation of challenges linked with climate change–related disturbances such as floods, forest fires, and drought.

Also, computer vision, when combined with machine learning, can control and monitor invasive species. They are often used in tracking and eliminating plant diseases, identifying the presence of invasive species. For example, AI and computer vision technology is applied to detect and uncover biodiversity changes by Blue River Technology. The Spatial Monitoring and Reporting Tool (SMART) is a technology-based tool to enable local populations to manage conservation areas by utilizing a digital platform in conjunction with conservation capacity-building efforts. However, given the importance of conservation technology for biodiversity conversation, we need to understand and explore some inadvertent consequences on nature and wildlife. For example, one could research the invasive nature of these technologies regarding how drones result in physiological and behavioural changes in nature species.

10.4.1.3 Harnessing Potential of Technology to Improve the Health of the Ocean

New approaches to sustainably manage and protect oceans are developed by AI techniques; some systems use machine learning and image analytics for species protection by identifying invasive species location. To track unreported, unregulated and illegal fishing, AI is also unlocking new solutions to prevent overfishing. To monitor illegal fishing activities, efforts are in progress to apply vessel algorithmic forms to automatic identification system (AIS) data from ships. According to European Commission research, robotic fish technology with AI capabilities can find potentially hazardous pollutants. To determine the best area for planting, data are processed to discover which species are best suited for the particular location. Moreover, to take thorough imagery of marine habitats, drones and machine learning algorithms are being developed.

The environmental engineering company Ocean Cleanup, based in the Netherlands, is using the digital technology to identify piles of garbage in the ocean. Using satellite images, combined with short-wave infrared sensors and remote sensing, locations of the densest plastics are identified and the plastics extracted from the oceans and rivers to reduce ocean pollution. In order to catalyze global efforts to protect the ocean's ecosystems and species, The Nature Conservancy organization is applying AI, electronic reporting, and electronic monitoring. These technologies can transform the way fisheries are managed by using on-board cameras and electronic monitoring to capture fishing activities. Similarly, fishermen can enter data into smart phones, tablets, and computers through electronic reporting and can communicate the information to seafood exporters, management agencies, and governments to enable them to make informed decisions.

10.4.1.4 Digital Technologies for Water Security, River Quality, and Drought Prediction

The supply of clean water and sanitation has been recognized by the UN as one of the Sustainable Development Goals, taking cognizance of the importance of water security for global development. In order to manage water resources efficiently, water vulnerabilities and availability must be continuously monitored. Digital technology capabilities if harnessed efficiently can increase global access to clean and safe drinking water. The accessibility of remote sensing technologies act as an important instrument for water resource overseers and researchers for observing water bodies, which was otherwise not possible by using ground-based monitoring.

AI and machine learning technologies, combined with satellite data, can track river and reservoir water quality, predict drought conditions, and forecast weather conditions, which in turn contributes towards informed policy decisions. For example, using machine technologies, Flo Technologies provides real-time data with water quality notification alerts to smartphones users. Similarly, by applying a data analytics platform, Water Smart Software offers the ability to spot anomalies in water meter data and check water flows. Blockchain technology is capable of tracking water tampering during the transfer process as well as issues of illegal extraction of water. Digital technologies can help analyze surface water conditions, analyze soil conditions, and help forecast weather patterns. Moreover, these technologies in combination with satellite imagery enables detection of underground leaks in portable water supply systems and a reduction in water loss. Mechanization of water usage control and monitoring can also lead to greater energy conservation. The gap between supply and demand of water resources due to excess exploitation or lack of rain can be bridged by using block chain technology. However, to realize the full potential of digital technologies, it's imperative that projections and forecasting generated by these technologies should provide useful insights to decision-making communities, researchers, and other stakeholders involved in water resources conservation.

10.4.1.5 Digital Technologies for Pollution Reduction and Air Quality Alerts

Air pollution is a critical global challenge having harmful health impacts. Over the past few decades, there has been growing emphasis on developing digital platforms to communicate air quality on a real-time basis to improve health outcomes and reduce pollution exposure. Many technologies such as machine learning algorithms and AI are used in predicting heat waves, climate change, airborne diseases, and various harmful pollutants. To diagnose, manage, and treat illnesses brought on by air pollution, many researchers have adopted AI techniques as an improved health-care decision support tool [34]. Many researchers have demonstrated that when it comes to prediction and forecasting, artificial technology is more accurate than conventional statistical methods.

In urban areas, the use of digital technologies when applied to cameras, radar sensors, and vehicles to optimize traffic flow can contribute towards improving air quality via its effect on stop-start driving and stationary vehicles. To improve efficiency and effectiveness of electric vehicles, AI is utilized to develop advanced and innovative battery design. Advances in monitoring of real-time air quality are also driven by digital technologies. For air pollution tracking, the company AirTick uses machine algorithms to analyze images across cities at low cost and smartphone cameras as an alternative for air pollution detection sensors. By using neural network algorithms to relate a pollutant with a source in each given environment, smart indoor air quality monitors provide real-time information and can detect air pollution sources. To meet individual needs, sensor-based air purifying techniques give personalized clean air levels.

In order to detect the sources of air pollution, UNDP, in partnership with other organizations, has developed the GeoAI platform, which is powered by mobile technology, satellite imagery, and artificial intelligence to track the location and improve environmental compliance of the brick manufacturing industry in India.

10.4.1.6 Weather and Disaster Resilience Action Areas

For the purpose of weather and disaster resilience, scientific communities and governments can monitor and forecast sea level changes, windstorms, floods, tremors, and many more potential natural disasters in real-time with thresholds for automated triggers that enable early evacuations with the help of predictive analytics powered by blockchain, advanced sensor platforms, IoT, drones, and AI. Image analytics and deep learning algorithms can use satellite images, seismic data, and building structural data to identify the areas of a city that will be most at risk for concerted disaster relief efforts for the purpose of real-time response planning. In response to enquiries, machine learning techniques and natural language processing are utilized to notify the public about disasters (e.g., Iowa Flood Information System (IFIS) Knowledge Engine). PetaBencana.id uses AI, various open-source sensors, and user reports from social media in Indonesia to map floods in real time in the nation's capital, Jakarta.

Leveraging the power of digital technology, UNDP has implemented a project on "accelerating disaster risk reduction and enhancing crisis response through digital solutions" in Philippines, Sri Lanka, Nepal, and Indonesia. This project contributes towards informed decision-making, planning, budgeting, and using disaster data to identify who is most affected by the disaster, which has three outcomes:

1. To reduce disaster risk and support recovery in countries, disaster data is digitalized and integrated by applying cloud-based technologies.
2. In order to increase disaster preparedness, targeted digital solutions and response of susceptible groups are developed and applied in the countries.
3. For disaster resilience and informed development planning, the capacities of subnational and national governments are strengthened through partnerships.

10.5 SCOPE FOR FUTURE RESEARCH

The same technology that facilitates protection of nature can also be applied for objectives that go against conservation goals. For example, technologies such as GPS can contribute towards resource depletion and intense fishing, and drones and camera traps could be used to enable illegal hunting. Many technologies increase the intensity of energy consumption and generate e-waste, which is mostly exported to developing counties. Moreover, environmental impacts of digital technology are hard to measure. The efficiency of using digital services increases due to technological improvements, which further increases energy consumption. Next, we will discuss some of the challenges associated with digital technology implementation for sustainability purpose and future research that should focus on overcoming these impediments.

1. As the use of digital technologies in day-to-day life becomes more prominent, there will be an upstream shifting of energy demand and related emissions from users to providers of digital infrastructure and services. While assessing the environmental impacts of digital technologies, effects of material basis, mining, production, and disposal of digital devices, and infrastructure development are often ignored, for example, loss in biodiversity, contamination of water or soil, and land-use changes. In order to have a comprehensive assessment of implications of digital technologies on sustainability, upstream shifting of environmental impacts must be taken into consideration.
2. The increase in demand for digital devices (mobile phones, computers, etc.) and high-tech gadgets have led to the emergence of "digital e-waste", which have serious environmental consequences. All forms of digital technologies contain rare earth elements such as minerals and metals, which are recycled

from e-waste and necessary for their functioning. The exposure to harmful materials such as palladium, cadmium, lead, and chromium poses serious health risks, including lung cancer and respiratory diseases, especially to vulnerable groups. The process of extraction causes water pollution with chemicals, and considerable land degradation and soil degradation, and entails cleaning operations. Recycling and disposing e-waste in an environmentally sustainable manner poses a big challenge.

3. The intensity of ultraviolet rays from the sun is most likely to increase due to depletion of the ozone layer, which in turn will hamper the electromagnetic frequency along which Wi-Fi travels. In coastal areas in the coming 15 years, the functioning of the internet may be disrupted due to damage to submarine cables on account of rising sea levels. Hence, climate change itself poses a challenge for the successful implementation of digital technology in the future.
4. The highly uneven access to digital knowledge, technologies, devices, data, and supporting infrastructure poses a challenge for significant operation and development of environmentally promising application. For example, a few "data-rich" corporations can lead to data-based market concentration, which in turn can have the disadvantage of sustainable applications and privilege the dissemination and development of detrimental innovations. Hence, right to use and access to data in line with common good has relevant sustainability concern. Moreover, legal arrangements and appropriate policies for equal access to digital technologies needs to be devised for proper environmental evaluation.
5. To protect against unethical use of data, discriminatory algorithms, and job displacement resulting from technology, responsible technology governance is needed. With increased dependency on the internet, cyberattacks are up and this needs to be addressed. The capacity of digital technology to enable automated analysis of data, and spatially and intensified distributed surveillance cause significant infringement of human rights. For example, the use of drones to collect data from a national park by wildlife authorities to abate poaching may violate tourists' privacy. Debate about issues such as whether people should consent to data collection and who should be permitted to deploy such devices (private and public organizations) are central among human rights organizations and of great relevance to sustainability or nature conservation.
6. Although digital technologies hold promising prospects for future sustainability, the digital divide could prohibit large-scale adoption. Digital technologies are expensive and out of reach of lower income, developing countries, which contribute most to climate change. Moreover, technical barriers, lack of expertise, and capacity gaps can hinder implementation of digital technologies for sustainability purposes, especially in low-income countries. For example, to generate analytical information, specialist knowledge is often required by GIS software.

10.6 CONCLUSION

Digital technology is expected to reduce carbon emissions worldwide by generating energy-saving benefits, developing transportation systems, smarter cities, next-generation distributed networks, and industrial processes. Emerging technologies such as robotics and automation, drones, blockchain, and AI contribute to the UN's Sustainable Development Goals by precisely monitoring and controlling complex processes, improving the ability to reduce pollution and enhancing energy efficiency. Because real-time monitoring transparency in supply chains will improve, the environment can be more easily investigated and protected.

This chapter highlights the research potential and future trends of digital technologies for sustainability purposes. It evaluates the implications of digital technology such as cloud computing, blockchain, IoT, big data, and AI on, for example, pollution reduction, sustainable farming practices, conservation of biodiversity, and natural disasters management. Using real-life cases, the chapter investigates how digital technologies can be a both obstacle and enabler to global sustainability, which will enable devising appropriate digitalization strategies geared towards the achievement of sustainability.

Digital technologies have the immense potential to unlock the solutions or way in which environmental problems are tackled. We have used real-life cases across relevant action areas pertaining to climate change, clean air, biodiversity and conservation, healthy oceans, water security, and weather and disaster resilience to demonstrate how each area stands to benefit from the application of these technologies.

AI algorithms, cloud computing, and big data analytics can be used to predict natural disasters such as floods, droughts, wildfires, and earthquakes, and preserve and monitor endangered species on land. Smart grids powered by AI can reduce GHG emissions by monitoring trends in energy consumption. Satellite imagery and data collection tools can monitor the pollution and health of ocean habitats and marine ecosystems, prevent overfishing, and thus ensuring ocean sustainability. IoT combined with smart surveillance and security systems can be used to monitor animal behaviours and prevent poaching practices by tracking endangered animals.

However, the same technology that facilitates protection of nature can also be applied for objectives that go against conservation goals. Many technologies increase the intensity of energy consumption and generates e-waste, which is mostly exported to developing counties. Moreover, environmental impacts of digital technology are hard to measure. The efficiency of using services increases due to technological improvements, which further increases energy consumption. The increase in demand for digital devices (mobile phones, computers, etc.) and high-tech gadgets have led to the emergence of "digital e-waste", which has serious environmental consequences.

Although digital technologies hold promising prospects for future sustainability, yet the digital divide could prohibit large-scale adoption. These technologies are expensive and out of reach of lower income, developing countries, which contribute most to climate change. Moreover, technical barriers, lack of expertise, and capacity gaps can hinder implementation of digital technologies for sustainability purposes, especially in low-income countries. Hence, future research should focus on making technologies affordable and facilitate large-scale adoption of sustainable digital technologies.

REFERENCES

1. L.S. Iyer and S. Venkatagiri,"Is digital good for providing water security? – A case study of India", *International Journal of Advanced Science and Technology*, Vol. 28, pp. 1074–1092, 2019, http://sersc.org/journals/index.php/IJAST/article/view/2062.
2. B. Ekholm and J. Rockstrom, "Digital technology can cut global emissions by 15%. Here is how", *In the Technological Transformation, The World Economic Forum*, January 2019, https://www.weforum.org/agenda/2019/01/why-digitalization-is-the-key-to-exponential-climate-action/
3. J.M. Müller, D. Kiel and K.I. Voigt, "What drives the implementation of Industry 4.0? The role of opportunities and challenges in the context of sustainability", Sustainability, Vol 10(1), p. 247, 2018, https://doi.org/10.3390/su10010247.
4. D.L. Northrup et al., "Novel technologies for emission reduction complement conservation agriculture to achieve negative emissions from row-crop production", *Proceedings of the National Academy of Sciences of the United States of America*, Vol. 118, p. 6118, June 2021, https://doi.org/10.1073/pnas.2022666118.
5. P. De Giovanni, "Smart supply chains with vendor managed inventory, coordination, and environmental performance", *European Journal of Operational Research*, Vol. 292(2), pp. 515–531, 2021, https://doi.org/10.1016/j.ejor.2020.10.049.
6. Environmental Defence Fund, "Fueling a digital methane future, The role of digital technologies in minimizing global oil and gas methane emissions", 2019, https://www.edf.org/sites/default/files/documents/Fueling%20a%20Digital%20Methane%20Future_FINAL.pdf.
7. C. Xiaohong et al., "Path of digital technology promoting realization of carbon goal in China's energy industry", *Bulletin of Chinese Academy of Sciences* (Chinese Version), Vol. 36, p. 1019, Article 6, 2021, https://doi.org/10.16418/j.issn.1000-3045.20210807004.
8. Z. Acs et al., "The evolution of the global digital platform economy: 1971–2021", *Small Business Economics*, Vol. 57, March 2021, https://doi.org/10.1007/s11187-021-00561-x.
9. V. Juyal, Nitin Pandey and Ravish Saggar, "Performance comparison of DTN multicasting routing algorithms-opportunities and challenges", *2017 International Conference on Intelligent Sustainable Systems (ICISS)*, Palladam, 2017, pp. 53–57. https://doi.org/10.1109/ISS1.2017.8389238.
10. V. Juyal, Nitin Pandey and Ravish Saggar, "An anatomy on routing in delay tolerant network", *2016 IEEE International Conference on Computational Intelligence and Computing Research (ICCIC)*, Chennai, 2016, pp. 1–4. https://doi.org/10.1109/ICCIC.2016.7919724.
11. V. Juyal, Nitin Pandey and Ravish Saggar, "Impact of varying buffer space for routing protocols in delay tolerant networks", *2016 International Conference on Communication and Signal Processing (ICCSP)*, Melmaruvathur, 2016, pp. 2152–2156. https://doi.org/10.1109/ICCSP.2016.7754562.
12. Capgemini, "Is enterprise it ready to become sustainable it ?" June 2021, https://www.capgemini.com/insights/expert-perspectives/is-enterprise-it-ready-to-become-sustainable-it/.
13. S. Ahmad, S. Mishra and V. Sharma, "Green computing for sustainable future technologies and its applications", in Grima, S., Sood, K. and Özen, E. (Eds.) *Contemporary Studies of Risks in Emerging Technology, Part A (Emerald Studies in Finance,*

Insurance, and Risk Management), Emerald Publishing Limited, Bingley, pp. 241–256, 2023. https://doi.org/10.1108/978-1-80455-562-020231016.
14. N. Pushpalatha, S. Jabeera, N. Hemalatha, V. Sharma, B. Balusamy and R. Yuvaraj, "A succinct summary of the solar MPPT utilizing a diverse optimizing compiler", *2022 5th International Conference on Contemporary Computing and Informatics (IC3I)*, Uttar Pradesh, India, 2022, pp. 1177–1181, https://doi.org/10.1109/IC3I56241.2022.10072844.
15. M.A. Ali, B. Balamurugan, R.K. Dhanaraj and V. Sharma, "IoT and blockchain based smart agriculture monitoring and intelligence security system", *2022 3rd International Conference on Computation, Automation and Knowledge Management (ICCAKM)*, Dubai, United Arab Emirates, 2022, pp. 1–7, https://doi.org/10.1109/ICCAKM54721.2022.9990243.
16. M. Koot and F. Wijnhoven, "Usage impact on data center electricity needs: A system dynamic forecasting model", *Applied Energy*, Vol. 291, p. 116798, 2021, https://doi.org/10.1016/j.apenergy.2021.116798.
17. Analysts, G.I.: I. Internet Data Centres - Global Market Trajectory & Analytics, September 2021, https://www.businesswire.com/news/home/20210903005160/en/Global-Internet-Data-Centers-Market-Report-2021-Market-to-Reach-143.4-Billion-by-2027---Explosion-of-Big-Data-Creates-Opportunities---ResearchAndMarkets.com.
18. K. Durgalakshmi, P. Anbarasu, V. Karpagam, A. Venkatesh, B. Kannapiran and V. Sharma, "Utilization of reduced switch components with different topologies in multi-level inverter for renewable energy applications-A detailed review", *2022 5th International Conference on Contemporary Computing and Informatics (IC3I)*, Uttar Pradesh, India, 2022, pp. 913–920, https://doi.org/10.1109/IC3I56241.2022.10073430.
19. S. Basheer, K.U. Singh, V. Sharma, S. Bhatia, N. Pande and A. Kumar, "A robust NIfTI image authentication framework to ensure reliable and safe diagnosis", *Peer J Computer Science*, Vol. 9, p. e1323, 2023.
20. J. Birat, "How to tell the story of change and transition of the energy, ecological and societal systems", *Matériaux & Techniques*, Vol. 108, p. 502, 2020, https://doi.org/10.1051/mattech/2021005.
21. A.K. Feroz, H. Zo and A. Chiravuri, "Digital transformation and environmental sustainability: A review and research agenda", *Sustainability*, Vol. 13, p. 1530, 2021, https://doi.org/10.3390/su13031530.
22. S.E. Bibri and J. Krogstie, "The core enabling technologies of big data analytics and context-aware computing for smart, sustainable cities: A review and synthesis", *Journal of Big Data*, Vol. 4, p. 38, 2017, https://doi.org/10.1186/s40537-017-0091-6.
23. Y.K. Dwivedi, "Climate change and COP26: Are digital technologies and information management part of the problem or the solution? An editorial reflection and call to action", *International Journal of Information Management*, Vol. 63, p. 102456, April 2022, https://doi.org/10.1016/j.ijinfomgt.2021.102456.
24. V. Sharma, B. Balusamy, J.J. Thomas and L.G. Atlas (Eds.), *Data Fabric Architectures: Web-Driven Applications*. Walter de Gruyter GmbH & Co KG, 2023.
25. M.A. Ali, B. Balamurugan, R.K. Dhanaraj and V. Sharma, "IoT and blockchain based smart agriculture monitoring and intelligence security system", *2022 3rd International Conference on Computation, Automation and Knowledge Management (ICCAKM)*, Dubai, United Arab Emirates, 2022, pp. 1–7, https://doi.org/10.1109/ICCAKM54721.2022.9990243.
26. World Economic Forum, *Innovation with a Purpose: The Role of Technology Innovation in Accelerating Food Systems Transformation*, McKinsey & Company, World Economic Forum, January 2018, https://www3.weforum.org/docs/WEF_Innovation_with_a_Purpose_VF-reduced.pdf.

27. W. Zhao et al., "Augmenting crop detection for precision agriculture with deep visual transfer learning—A case study of bale detection", *Remote Sensing*, Vol. 13, p. 23, 2021, https://doi.org/10.3390/rs13010023.
28. How digital technology and innovation can help protect the planet, United Nations Environment Programme, August 2022, https://www.unep.org/news-and-stories/story/how-digital-technology-and-innovation-can-help-protect-planet.
29. A. Singh, M.A. Ali, B. Balamurugan and V. Sharma, "Blockchain: tool for controlling ransomware through pre-encryption and post-encryption behavior", *2022 Fifth International Conference on Computational Intelligence and Communication Technologies (CCICT)*, 2022, pp. 584–589, https://doi.org/10.1109/CCiCT56684.2022.00107.
30. PwC, Fourth industrial revolution for the earth: Harnessing artificial intelligence for earth, January 2018, https://www3.weforum.org/docs/Harnessing_Artificial_Intelligence_for_the_Earth_report_2018.pdf.
31. G. Bressanelli, F. Adrodegari, M. Perona and N. Saccani, "Exploring how usage-focused business models enable circular economy through digital technologies", *Sustainability*, Vol. 10, p. 639, 2018, https://doi.org/10.3390/su10030639.
32. PwC, "Building block(chain)s for a better planet, fourth industrial revolution for the earth series", 2018, https://www.pwc.com/gx/en/sustainability/assets/blockchain-for-a-better-planet.pdf.
33. K. Fritzsche, S. Niehoff and B. Grischa, "Industry 4.0 and climate change — Exploring the science-policy gap", *Sustainability*, Vol. 10, p. 4511, 2018, https://doi.org/10.3390/su10124511.
34. A. Masood and K. Ahmad, "A review on emerging artificial intelligence (AI) techniques for air pollution forecasting: Fundamentals, application and performance", *Journal of Cleaner Production*, Vol. 322, p. 129072, 2021, https://doi.org/10.1016/j.jclepro.2021.129072.
35. V. Juyal, Nitin Pandey and Ravish Saggar, "Opportunistic message forwarding in self organized cluster based DTN", *2017 International Conference on Infocom Technologies and Unmanned Systems (Trends and Future Directions) (ICTUS)*, Dubai, 2017, pp. 497–502. https://doi.org/10.1109/ICTUS.2017.8286060.

INDEX

2030 Agenda of Sustainable Development, 2, 15
 5 P's, 3, 4

A

Absorber–tube connection model, 37
Active solar power, 137
Advanced configuration and power interface (ACPI), 113, 122
Advanced Energy Initiative, 28
Agricultural by-products and wastes, 139
Agricultural sector, 28
Agricultural systems, 31
Agriculture, 28
Agrifood chain, 28
Air–air heat exchanger (AAHE), 139
Air pollution, 197
AirTick company, 197
Alternating current, 43
Anaerobic digestion (AD) procedure, 46
Apache Spark, 84
Apple, 61
Artificial brainpower, 76
Artificial intelligence (AI), 21
 in O&G industry, 76–77
 positive and negative impacts on sustainable development, 22–23
 sustainable digital technologies, 52–53
Artificial Neural Network (ANNs), 76, 77
Asia-Link initiative, 136
Aspirational Districts Programme, 128
Automatic identification system (AIS), 196
Automobile firms, 180
Autonomous vehicles (AVs), 195
Azure IoT Hub, 93

B

Basic themes, 169, 172, 175
Battery-based energy storage components, 40
Biblioshiny platform, 164
Big data, 52, 55–57
 in O&G industry, 81–82
  characteristics, 81
  downstream O&G operations, 83–84
  upstream O&G operations, 82–83
Biodegradable materials, 138
Biological oxygen demand (BOD), 100
Biopharmaceutical businesses, 61
Blockchain in O&G industry, 71–72
 opportunities and applications, 73
  discernment of well abandonment and restoration, 76
  following and tracing of O&G shipments, 74
  functional and compliance audit, 75
  gathering and disposing of waste at oil refinery stations, 76
  improving billing and payments, 75
  overseeing life cycle O&G assets, 75
  robotizing leakage detection of gas pipeline, 73–74
  robotizing O&G production, 74
 varieties, 72–73
Blockchain technology, 58–59, 73, 76, 162, 179, 194, 196
Blue River Technology, 195
Bridge rectifier, 43
Brundtland Commission, 2
Business-to-consumer (B2C) standpoint, 79

C

Cape Town Action Plan for Sustainable Development Goals, 11
 strategic action areas of, 12
Capital costs, 145
Centre for Research on Energy and Clean Air, 152
Circular economy (CE), 20, 162–163
Cleaner energy technology, 146
Cleaning, 36
Clean Out Kit, 180
Climate Savers Computing Initiative (CSCI), 127
Closed-loop reservoir management, 83
Cloud-based system, 54
Cloud computing, 52, 53–55
 in O&G industry, 80–81
Cold side heat exchanger (CSHE), 40
Coliform, 101
Collaboration, 177
Company production line, 54
Compound parabolic concentrators (CPCs), 37
Computer vision, 195
COMSOL Multiphysics simulation software, 38
COMSOL receiver model, 38
Consortium/united blockchains, 72
Consultative Group on International Agricultural Research (CGIAR), 28
Contaminated ocean water on ecosystem, 93
 architecture, 93, 99–101

dependency of human and ocean health on each other, 92
dissolved oxygen limits for aquatic life, 99
experimental setup
analysis of results, 105–106
decision forest algorithm, 104
Indian water quality data, 101
support vector machines (SVMs), 104–105
two-class neural network component, 103
literature review, 94–98
pH ranges of surface ocean water, 100
Cooling-as-a-service business model, 180
Cool roofs, 138–139
CoreCentric Solutions, 181
Coronavirus SARS-CoV-2, 36
Corporate social responsibility (CSR), 115
COVID-19, 14, 36, 60, 149, 150
impacts, 150–155, 157, 158, 187
Cracked gas compressor (CGC), 83, 84
Cross-industry collaborations, 177
Cyber-physical systems (CPS), 57, 58
Cybersecurity risks, 54

D

Daylighting design principles, 141
DC–DC converter, 40
block diagram, 41
Decision forest algorithm, 104
Decision-making, 19
Declining/emerging themes, 169, 172, 175
Derwent Innovation Index (DII) database, 32
Development Data Council, 11
Digitalization, 11
Digital sustainability, 11–13, 28
innovation in sanitization, 34–36
need of patents and innovations, 28–29
photovoltaic technologies, mapping from patent analysis
hierarchical tree, 34
patent classification, 33–34
photovoltaic cell technology, 33
Digital technologies, 188–190, 200
applications
biodiversity and conservation, 195
climate change challenges, 194–195
harnessing potential of technology to improve health of ocean, 195–196
pollution reduction and air quality alerts, 197
water security, river quality, and drought prediction, 196–197
weather and disaster resilience action areas, 197–198
in Industry 4.0
artificial intelligence (AI), 52–53
big data, 55–57
blockchain, 58–59
cloud computing, 53–55
healthcare industry, 61–62
Internet of Things, 57–58
and innovations, 191–193
harnessing power of data, 193
scope for future research, 198–199
sustainability in business, 194
trends in, 190–191
Disinfection, 36
Dissolved oxygen (DO), 93
Downstream O&G operations
big data in O&G transportation, 84
big data in refining, 83–84
use of big data by health and safety executive (HSE), 84
Drill-string dynamics, 83
DSM–Niaga, 178
Dual plumbing, 140

E

Earth Solidarity Tax, 151
EatCloud platform, 188
Eco-effectivity, 163
Eco-efficiency, 163
Eco-friendly buildings
power savings, 141–142
renewable energy technologies and longterm viability, 146
Eco-friendly computer, 110
Eco-ICT, 112
Ecological disbalance, 14
Ecological sustainability, 11
Economic expansion, 150
Economic sustainability, 9–10, 19
Ecosystem Integrity Fund (EIF), 31
schematic vision of, 32
Electric renewable identification number (eRIN), 45, 46
Electric submersible pumps (ESPs), 83
Electric tractors, 31
Electric vehicles (EV), 31, 46
Electrochromic smart glass, 139–140
Electronic auxiliary equipment, 31
Electronic Product Environmental Assessment Tool (EPEAT), 121
Energy Policy Act of 2005, 45
Energy sector, 192
Energy Star programme, 114
Enhanced oil recovery (EOR) projects, 83
Environmental degradation, 28
Environmental sustainability, 19
European Commission, 136, 143, 196

Extended Producer Responsibility scheme, 179
Extreme gradient boosting (XGBoost), 84

F

Fats, oils, and greases (FOGs), 46
Flat/sloping roofs, 143
Flexible Plastic Fund initiative, 177
Flo Technologies, 196
FLOTHERM, 40
Frameless lenses, 144
Freshwater Ecosystem Explorer, 193
Fresnel lenses, 144
Full-bridge rectifier, 43

G

Game-theoretical model, 163
GeoAI platform, 197
GIS software, 199
Global Environment Monitoring System for Air (GEMS Air), 193
Globalization, 1
Global tourism industry, 151
Google Cloud, 84
Green building, 136–137
  eco-friendly buildings
    power savings, 141–142
    renewable energy technologies and longterm viability, 146
  energy-saving and eco-friendly technologies raising standard for green design, 142–144
    atriums construction using Fresnel lenses, 144
    financial impact of renewable energy technologies, 145–146
    systems combining photovoltaic and thermal (PV/T) energy, 144
    wind/solar systems integration, 144–145
  impact of renewable energy sources and techniques, 137
    biodegradable materials, 138
    cool roofs, 138–139
    electrochromic smart glass, 139–140
    green insulation, 138
    low/no energy consumption designing, 139
    power-sufficient structures, 140
    rammed earth brick, 140–141
    smart appliances, 138
    solar energy, 137–138
    sustainable indoor environment technologies, 140
    sustainable resource sourcing, 139
    systems for efficient use of water, 140
Green computing, 109, 110, 131–132
  advantages, 117
  approaches, 114
    green design, 113
    green disposal, 114
    green manufacturing, 113
    green use, 114
  benefits, 116
    improved social and corporate image, 115
    meeting compliance and regulatory requirements, 116
    resource usage, 115
    risk management, 116
    saving money, 115
    sustainability of environment, 115
  disadvantages, 118
  effects of IT devices and wastes on health, 120
  future, 129, 131
  history, 114–115
  implementation procedure, 118–119
  implementation strategies
    calculating present carbon footprints of IT components, 125
    creating strategy for long-term green computing, 124–125
    electronics disposal in useful and eco-friendly way, 125
    enhancing business performance with sustainable IT practices, 126
    saving energy, 125–126
    utilization of powerful computer applications, 125
  industry and government schemes
    government initiatives, 126–127
    industry initiatives, 127
    Niti Aayog initiative, 127–128
  limitations, 117–118
  literature review, 111–112
  necessity, 112–114
  need of study, 110–111
  objective of study, 111
  policy implementation in three leading countries, 129
  practices, 121
    algorithm efficiency, 122
    display, 124
    implementation of enhanced data centre cooling methods, 122
    power management, 122
    power supply, 122, 123
    storage, 124
    virtualization, 122
  solutions
    creating long-term green computing strategies, 119

lowering use of paper, 121
power-saving measures, 121
purchasing environmentally friendly and ethical goods, 121
reusing and recycling, 119, 121
sustainable development in post-COVID era, 128–129
Green Computing Impact Organization Inc. (GCIO), 127
Green electricity production, 47
Green Electronics Council, 127
Green Grid, 127
Greenhouse gas emissions (GHG), 28, 31, 45, 112, 153, 188
Green insulation, 138
Green IT, 129
contribution to sustainable development, 130–131
Green technology, 138

H

Hackers, 54
Hadoop framework, 84
Hanover Fair, 70
Harmful algal bloom (HAB)-produced toxins, 93
Health and safety executive (HSE), 84
Healthcare industry, 61–62
Heating Ventilation Air Conditioning (HVAC), 110
Heat transfer medium, 29
Heliostats, 29, 37
10's multiaxis motorized system, 30
Hot side heat exchanger (HSHE), 40
Human activities, 13, 23
Human resources, 60
HVAC systems, 139
Hydroelectricity, 142
Hyperledger Fabric platform, 72

I

IKEA, 177, 178, 181
Importance–performance map analysis (IMPA), 77
Indian education technology (edtech), 154–155
blended online mentoring programmes, 156
COVID-19 impact on Indian education system, 155
future of edtech sector, 155–156
Indian edtech industry, 155
prospects for future, 156
as sustainable business model, 157–158
Indoor air quality, 140
Industrial IoT (IIoT), 78
Industry 4.0, 19, 52, 189; *see also* Digital technologies, in Industry 4.0; Oil and gas (O&G) industry
benefits, 63–64
challenges, 62–63
in different sectors, 59–61
technologies in, 70–71
Industry manufacturers, 56
Infrastructure as a service (IaaS), 81
Institute for Global Environmental Strategies (IGES), 151
Insulation, 138
Integrated asset modelling (IAM), 83
Intel, 56
Intellectual property (IP) domain, 59
Intelligent traffic junction, 44
Interactive visual analytics, 83
Intergovernmental Panel on Climate Change (IPCC), 4, 28
Intermediate solar irradiance, 144
International Methane Emissions Observatory (IMEO), 193
International Union of Nature and Natural Resources (IUCN), 4
Internet of Things (IoT), 34, 54, 55, 57–58, 93
in O&G industry, 77
architecture, 80
environmental value, 79
firm size, 78
information intensity, 78–79
management support, 78
security concerns, 78
trading partner pressure, 78
Invisible non-productive time (INPT), 83

K

Kaer, 180

L

Large car and truck carriers (LCTCs), 84
Lightning-as-a-service business model, 179
Li-ion reversible battery technologies, 43
Linear Fresnel reflector (LFR), 37
Liquefied petroleum gas (LNG), 74
Logging while drilling (LWD), 81
Low-Carbon Gasoline Standard (LCFS), 45
Low/no energy consumption designing, 139
Low-power wide-area (LPWA), 79

M

MacBooks, 118
Machine learning (ML), 54, 106
Machine-to-machine (M2M) communication, 77

Manufacturing execution system (MES), 84–85
  milestones, 85
Massive open online courses (MOOCs), 158
Measurement and verification (M&V), 84
Measuring while drilling (MWD), 81
Methane, 193
Micro-mirror solar collector, 36–38
Millennium Declaration, 3
Mobility sector, 192
Monte Carlo ray tracing (MCRT) procedure, 38
Motor themes, 169, 174
Multi-layer perceptron (MLP) neural networks, 84
Multiple sourcing, 177
Multiple worldwide solar integration systems, 143

N

National Oceanic and Atmospheric Administration (NOAA), 100
National Patent Development Strategy, 34
National Sustainable Development Strategy, 3
Natural materials, 137, 138
Nature Conservancy organization, 196
NEOMA Business School, 153
Niche themes, 169, 172, 174
Niti Aayog initiative, 127–128
Non-profit Green Electronics Council, 121

O

Ocean acidification, 100
Ocean Cleanup, 196
Official development assistance (ODA), 3
Oil and gas (O&G) industry, 70
  artificial intelligence, 76–77
  big data, 81–82
    characteristics, 81
    downstream O&G operations, 83–84
    upstream O&G operations, 82–83
  blockchain, 71–72
    opportunities and applications, 73
    varieties, 72–73
  cloud computing, 80–81
  Internet of Things (IOT), 77
    architecture, 80
    environmental value, 79
    firm size, 78
    information intensity, 78–79
    management support, 78
    security concerns, 78
    trading partner pressure, 78
  manufacturing execution system (MES), 84–85
    milestones, 85
On-demand production, 176
On-site energy production, roadmap of, 31
Operating systems (OS), 112–113
Optical instruments, 144
"Our Common Future" report, 2

P

Parabolic trough concentrator (PTC), 37
Parkinson's disease, 61
Partnership in Statistics for Development in the 21st century (PARIS21), 11
Passive building, 137
Passive solar power, 138
Patent data, 33
people, planet, prosperity, peace, and partnerships (5 P's), 3, 4
Photovoltaic cell technology, 33
Photovoltaic technologies, 32–33
  hierarchical tree, 34
  patent classification, 33–34
  photovoltaic cell technology, 33
Photovoltaic/thermal (PV/T) solutions, 144
Piezoelectric based shoe charger, 42–43
  block diagram, 44
  circuit diagram of bridge rectifier, 43
  diode, 42
Piezoelectric device, 42
Piston spring, 44
Platform as a service (PaaS), 81
Potential of hydrogen (*pH*) of water, 93
Poverty, 3, 7, 9, 10, 13–15, 21, 130, 143
Power-sufficient structures, 140
Power supply units (PSUs), 122
Precision agriculture, 192
Private blockchains, 72
Product portfolio, 176
Public blockchains, 72
Public participation entrepreneurship, 154

R

Radio frequency identification, 179
Rammed earth brick, 140–141
Rapid industrialization, 13
Reduce–reuse–recycle concept, 163
Regional circulating and ecological spheres (Regional-CES), 150
Renewable energy systems, 31, 142
  autonomous agricultural off-road vehicles, 31–32
  of waste heat recovery system for automobiles
    effects of temperature contact, 39–40
    system structure, 38–39
    TEG evaluation, 39
Renewable energy technologies
  financial impact of, 145–146

and longterm viability of eco-friendly buildings, 146
Renewable Identification Numbers (RINs), 45
Renewable sources of energy, 137
Resale-as-a-service business model, 180
Reverse logistics, 175
RFS-compliant bioenergy supply, 45
Roadways, 44
Robotics technology, 60
Robotic sterilization equipment, 36

S

Salinity, 100
Sanitization procedure
flow diagrams of, 35
innovation in, 34, 36
Seismic data analysis, 82
Shale operators, 83
Signify, 179
Smart appliances, 138
Smart building energy management, 142
Smart grids, 194, 195
Smart traffic roads junction energy exploration, electric energy generation, 44–45
Smartwatches, 61
Social sustainability, 10, 19
Software as a service (SaaS), 81
Solar cell contractor system, 30
Solar energy, 29, 137–139
Solar receiver, 30
Solar thermal systems, 37
Solar umbrella based on Peltier effect, 40–41
block diagram, 40
proposed design, 42
Solstice, 38
Spatial Monitoring and Reporting Tool (SMART), 195
Spillover effect, 189
Standard economic theory, 136
Sterilization, 36
Strasbourg Agreement of 1971, 33
Supervised learning technique, 103
Support vector machines (SVMs), 104–105
Sustainability, defined, 2
Sustainable building practices, 138
Sustainable business models (SBMs), 20
Sustainable construction toolbox, 140
Sustainable development, 1
components of, 8–9
ecological sustainability, 11
economic sustainability, 9–10
social sustainability, 10
conceptual relationship between sustainability and, 9
core concepts, 3
critical factors
climate change, 14
degradation of cultivated land, 14
ecological disbalance, 14
global warming and climatic change, 13
overexploitation of natural resources, 13
poverty, 13–14
scarcity of resources, 15
definitions, 2, 4, 5–7
deployment of digital resources across Asia Pacific, 13
digital sustainability, 11–13
gap in funding and statistics, 7
history of, 5
holistic impacts of digitalization and sustainability, 8
implementation progress of, 11
and Industry 4.0, 19, 20
in information technology (IT) sector, 19–21
literature review, 7–8
role of artificial intelligence in, 21–23
significance of, 2
sustainable development goals (SDGs), 15–19
themes and initiatives of technology companies, 20
Sustainable Development Goals (SDGs), 2, 7, 11, 15–19, 21, 163, 172, 190, 191, 193, 196, 200
Sustainable digitization, 188
Sustainable indoor environment technologies, 140
Sustainable interior innovations, 140
Sustainable practices in post-pandemic world, 150
COVID-19 impacts, 150–153
embracing green capitalism, 154
improving employee morale and output, 153–154
rise of Indian education technology, 154–155
blended online mentoring programmes, 156
COVID-19 impact on Indian education system, 155
edtech companies as sustainable business model, 157–158
future of edtech sector, 155–156
Indian edtech industry, 155
prospects for future, 156
Sustainable resource sourcing, 139
Sustainable supply chains in circular economy, 162, 163
descriptive analysis
annual scientific production, 166
author's network, 166, 168
countries network, 167, 169
journals list, 167, 170
most cited countries, 166, 167

most relevant affiliations, 166, 167
thematic analysis, 169, 172, 174
thematic evolution, 172, 174–176
word cloud, 167–169, 171–173
limitation and future scope, 181–182
research methodology, 164–165
strategies, 176–177
altering business models, 179–180
collaborating among partners, 177
enabling circular product designing, 178
financial incentives to support and make affordable circular economy practices, 178–179
leveraging technologies and data support to drive circularity, 179
paying attention to packaging of products, 178
reduce, reuse, and recycle products, 180–181

T

Tactical instrumentalists, 115
Take–make–dispose concept, 163
Thermoelectric generators (TEGs), 38
Thermoelectric module, 40
ThredUP, 180
Tractor manufacturers, 31
Traditional building practices, 138
Triple bottom line (TBL), 19
T-test method, 84
Two-class neural network component, 103

U

Ultra-thin thermoelectric cooler modules, 41
UN Business and Human Security Initiative, 152
Unglazed photovoltaic systems, 144
Unglazed solar thermal collectors, 144
UN Global E-Waste Monitor, 180
United Nations Biodiversity Lab 2.0, 193
United Nations (UN) Commission of Environment and Development, 1
United Nations Conference on Trade and Development, 162
United Nations Department of Economic and Social Affairs (UN DESA), 187
United Nations Development Programme (UNDP), 128, 197, 198
United Nations Environment Programme (UNEP), 193
United Nations General Assembly, 3
United Nations Statistics Division (UNSD), 11
United Nations Sustainable Development Goals, 127, 163
Unsustainable manufacturing, 150
Upstream O&G operations
big data in drilling, 83
big data in exploration, 82
big data in production engineering, 83
big data in reservoir engineering, 83
US Environmental Protection Agency (EPA), 115
US Renewable Fuel Standard (RFS), 45
business difficulties, 46
technology, 46

V

Virtual computers, 113, 122
VOSviewer software, 164–166, 168, 182

W

Waste heat recovery system for automobiles
effects of temperature contact, 39–40
system structure, 38–39
TEG evaluation, 39
Water (Prevention and Control) Act, 1974, 10
Water–air heat exchanger (WAHE), 139
Water jacket for solid particle solar receiver, 29–30
Water Quality Index (WQI), 101
Water Smart Software, 196
Web Driven technologies framework, 163
Wind and solar energy systems, 144–145
Wind turbines (WTs), 145
World Bank report, 8
World Conservation Strategy report, 4
World Economic Forum, 188, 191, 194
World Energy Strategy, 28
World Intellectual Property Organization, 33
World Wide Web, 52, 57

Z

Zipped assembly points, 40